BERGSON

The Arguments of the Philosophers

EDITED BY TED HONDERICH

Grote Professor of the Philosophy of Mind and Logic,
University College, London

The purpose of this series is to provide a contemporary assessment and history of the entire course of philosophical thought. Each book constitutes a detailed, critical introduction to the work of a philosopher of major influence and significance.

Already published in the series:

*Augustine	Christopher Kirwan
*J. L. Austin	Geoffrey Warnock
Ayer	John Foster
*Bentham	Ross Harrison
*Bergson	A. R. Lacey
*Berkeley	George Pitcher
Butler	Terence Penelhum
*Descartes	Margaret Dauler Wilson
*Dewey	J. E. Tiles
Gottlob Frege	Hans Sluga
Hegel	M. J. Inwood
*Hobbes	Tom Sorell
*Hume	Barry Stroud
*Husserl	David Bell
William James	Graham Bird
*Kant	Ralph C. S. Walker
*Kierkegaard	Alastair Hannay
*Locke	Michael Ayers
Karl Marx	Allen Wood
Meinong	Reinhart Grossman
*John Stuart Mill	John Skorupski
*G. E. Moore	Tom Baldwin
*Nietzsche	Richard Schacht
Peirce	Christopher Hookway
*Plato	J. C. B. Gosling
*Karl Popper	Anthony O'Hear
*The Presocratic Philosophers	Jonathan Barnes
*Thomas Reid	Keith Lehrer
*Russell	R. M. Sainsbury
Santayana	Timothy Sprigge
*Sartre	Peter Caws
*Schopenhauer	D. W. Hamlyn
*Socrates	Gerasimos Xenophon Santas
Spinoza	R. J. Delahunty
*Wittgenstein	Robert J. Fogelin

*available in paperback

BERGSON

A. R. Lacey

London and New York

First published 1989
by Routledge
11 New Fetter Lane, London EC4P 4EE

Simultaneously published in the USA and Canada
by Routledge Inc.
29 West 35th Street, New York, NY 10001

First published in paperback 1993

Typeset by Gilfillan Ltd, Mitcham, Surrey

Printed in Great Britain by T. J. Press (Padstow) Ltd.,
Padstow, Cornwall

British Library Cataloguing in Publication Data

A catalogue record for this book is available from the British Library

Library of Congress Cataloging in Publication Data

Has been applied for

ISBN 0–415–08763–5

Contents

Preface

I have tried in this book, in accordance with the title of the series, to state and examine Bergson's main arguments on their merits. So far as Bergson can be classified at all it would be as a 'process philosopher', along with Heraclitus, the Stoics, Hegel, and Whitehead. His nearest philosophical congeners are Whitehead himself, with whom he shares a mathematical background and an interest in modern physics, and William James, with whom he shares a strong pragmatist element and whom he regarded as a friend and ally. But I have not attempted any foray into the history of ideas, apart from the odd stray remark. I have approached the subject from the broadly 'analytical' standpoint prevalent in current English-speaking philosophy. This is by no means identical with Bergson's own standpoint, but I have tried to steer between unsympathetic rejection and uncritical overestimation.

As on previous occasions, I am greatly indebted to Professor Ted Honderich for his sympathetic encouragement as series editor, and to Dr John Watling for many philosophical discussions on topics relevant to those here treated. Dr Donald Gillies and Dr Harmke Kamminga of the Department of the History and Philosophy of Science at King's College read and gave me valuable comments on a draft of chapter 7. Parts of chapter 2 were read to a seminar in the Department and I am grateful for comments made there and especially for a prolonged later discussion with Mr Barrie Tonkinson. Dr Gillies, and Professor Roy Davies of Leicester University Mathematics Department, offered me considerable help on the end of chapter 6. Dr Mary Pickersgill and Dr Elizabeth Valentine of the Psychology Department of Bedford College gave me some valuable references in connexion with chapter 5. I hope the text and bibliography will make clear my debts to written sources though I cannot leave unmentioned the immense labour of love represented by P.A.Y. Gunter's Bergson bibliography (see my bibliography below). I am grateful to my colleagues in the King's College Philosophy Department for allowing me a sabbatical term at a helpful time and also for comments on parts of chapter 1 at a staff seminar. In this time of financial stringency I would like to say that any merits this book may have would never have reached the light of day without the excellent resources of the University of London library, whose philosophy librarian, Mrs Margaret Blackburn, has also been most helpful to me. Finally, I am grateful to Miss Kendall Anderson for her prompt and efficient production of a disc out of my minuscule handwriting and to Mrs Joan Waxman and Mrs Elizabeth Betts for a good deal of secretarial help of one kind or another at various stages.

List of Abbreviations

Bergson's works are abbreviated as follows:
(For details see bibliography)

CE	*Creative Evolution*
CM	*The Creative Mind*
L	*Laughter*
M	*Mélanges*
ME	*Mind-Energy*
MM	*Matter and Memory*
MR	*The Two Sources of Morality and Religion*
Q	*Oeuvres*
TF	*Time and Free Will*

I

Extensity and intensity

1. Introduction

Bergson was sometimes known as an idealist, though this is not a label that he would have applied to himself. In fact towards the end of his life he emphatically repudiated it in favour of realism, and the most radical realism at that (M, 1520 – 1, written in 1935). But his philosophy is built up on the notion of 'images', a term that has an idealist ring, and it could certainly be called a very 'experience-centred' philosophy. This philosophy of images comes to the fore chiefly in his second main book, *Matter and Memory*. In his first, *Time and Free Will*, he is concerned, at least on the face of it, with features of the objective world, mainly time and causation. Even here, though, the approach is through experience, as the book's original title shows: essay on the immediate data of consciousness. We will see later something of what 'immediate' means.

In the preface to *Time and Free Will* Bergson says that his main subject is free will, which the third chapter deals with, but he leads up to it with two introductory chapters on intensity and duration. The second of these has an importance in his philosophy as a whole that belies the label 'introductory', as he agrees in his résumé at TF, 226. But the first, though added as an afterthought (Q, 1542), contains what might well be called the germ of the whole system.

Magnitudes are often divided into two sorts, extensive and intensive. Extensive magnitudes are ordinary magnitudes, which allow of both comparison and measurement. Intensive magnitudes allow of comparison but not of measurement. This is because they do not admit of units which can be added together. Length is clearly an extensive magnitude. Happiness is an intensive magnitude if people can be called happier or less happy than other people, but not twice as happy. There are no units of happiness. But not all cases are so easy to classify. Sometimes units can be found but only if they are constructed from units proper to another magnitude. Miles per hour, as units of speed, are constructed from miles and hours. A body's temperature is the average speed of its molecules, and we measure this in ordinary cases by using its effect on the length of a tube of mercury. The quantity of heat in a body is the product of its temperature and its mass, itself a basic extensive magnitude. But how about the sensation of heat? We normally say we feel one thing as hotter than another, or that we feel hotter than we did yesterday. We wouldn't claim to feel one thing as being twice as hot as another, or ten per cent hotter than another, but would such claims make sense? Is it just that our senses are too rough and ready for such judgements? We must distinguish feeling one thing as ten per cent hotter than another, where we are meaning to talk of the sensations themselves which the things give us, from feeling *that* it is ten per cent hotter or feeling it as if it were ten per cent hotter (compare 'seeing as'), where in the first of these latter cases we are making a judgement and in the second case doing something parasitic on the making of judgements. These last two cases presuppose a notion of temperature independent of sensation, but not all feeling is like this. Men no doubt made comparisons of temperature long before they had heard of molecular motion, or had any other notion of objective temperature.

Bergson denies the whole notion of intensive magnitudes, which he sees as an attempt to treat qualities as quantities. He takes various examples of phenomena commonly regarded as quantitative and argues that they are really qualitative, the examples being various kinds of conscious states. The basic idea of his criticism comes right at the beginning in the first few pages of *Time and Free Will*. In cases where we can unambiguously say that one thing is greater than another, as we do with numbers or bodies, we do so because we can regard the smaller as contained in the larger, which goes beyond it. But, if we regard, say, one sensation as more intense or greater in degree than another, where does containment come in? How can we regard sensations as superposed on one another? We might try supposing that we could experience a more intense sensation by first experiencing a less intense degree of it, and then going beyond that. But this won't serve as an explanation unless we can justify the claim that the second stage does constitute going beyond the first rather than falling back from it. We have not got the less intense stage still there so that we can directly experience the second stage as adding to it. In an important remark Bergson explicitly links the extensive/intensive distinction with the extended/unextended one (TF, 3), a point which foreshadows the importance he will give to the notion of space.

2. The three main cases

However, common sense and the philosophers alike treat intensity as a magnitude and as something extended and we must give some account of this (TF, 3). Bergson first makes some preliminary remarks and then makes two distinctions, first between certain psychic states which seem relatively independent of external causes or physiological accompaniments and those which do not and then among these latter between sensations of effort of various kinds and pure sensation. (These three cases occupy respectively TF, 8 – 19, 20 – 31, 31ff.) The preliminary remarks are to the effect that it is no use trying to account for different degrees of intensity in sensations etc. by reference to different numbers either of causes of the sensations or of elementary parts of physical phenomena underlying them. The thing we are trying to explain, the apparent intensity of the sensation, is a property of the sensation itself, which we are aware of quite independently of any awareness we have of the cause or the accompaniments, as we saw in the case of temperature above. When we hear a loud sound the amplitude of the relevant vibrations in the air may be large and so may that of some movements in the brain, but this is not what we are referring to by using the term 'loud'; and the intensity of a tickle may decrease when the friction that causes it increases. Bergson does not consider what we now know as the identity theory of mental events, but one can imagine him saying of it in modern terms that it makes the identity contingent and therefore cannot use the amplitude of the vibrations to give the *meaning* of 'loud'. This is surely right.

The first of the three cases covers states like 'deep joy or sorrow, a reflective passion or an aesthetic emotion' (TF, 7). These seem not to involve 'the perception of a movement or of an external object' (TF, 7). They are simple in so far as they involve no extensive element, and so pure intensity should be more easily definable here. One might add that in another way they are also complex, as Bergson in fact brings out in his treatment of them. Though they may be relatively isolable from external causes or physiological accompaniments they are hardly pure states in themselves. They tend rather to 'tinge' other features of our mental life, to use Bergson's own term ('teigner'). He does not deny that they have an identity of their own, but he regards their intensity as the extent to which they do this tingeing, or the extent to which they absorb our whole personality. If, say, deep joy is not, or is not entirely, a simple state of mind, as he says, then it may not be surprising if it cannot have degrees of intensity, at least in a direct sense. To this extent the case is not a clearcut one. He does not deny that joy can be called deeper or less deep – to deny this would be to deny the phenomenon to be explained. But, if we say the difference in depth is really a difference in extent or in preoccupation, are we saying there is no difference in depth there but only a difference in extent, or that the difference in depth is real enough but is just a difference in extent? What precisely is the mistake we are to avoid? I think Bergson might say that either of these two ways of putting it could express the point, which is that there is nothing there that does differ in degree but does not differ in extent. Just because the case is not very clearcut it is hard to assess. For the moment I will make just two points. First we shall have to assume that we cannot be preoccupied by emotions, and our lives cannot be coloured by them, except in proportion as we regard them as intense; we cannot be preoccupied by emotions we regard as mild or minor – but this is different from saying we cannot be preoccupied by emotions we think ought to be mild, or which are concerned with considerations we judge to be minor. Second we face the question what is involved in there being different degrees of preoccupation, as there presumably are in some sense, and the question whether features of our life can be 'tinged' with joy etc. to different degrees, and if so what this involves.

In the second and third of the three cases the sensation is more closely linked to the external cause. Muscular effort might seem to be *par excellence* a case where sensation varies in degree: surely the greater the effort the more intense the sensation. Bergson claims, however, that the sensation in fact accompanies and depends upon certain muscular movements, and quotes empirical evidence involving paralysis to show that whenever the sensation of effort occurs certain muscles have actually moved, though these may not be the same muscles as the subject was trying to move. When the muscles cease to move the consciousness of effort vanishes (TF, 22, 23). In fact without going as far as a laboratory one can simply try this home experiment: can we experience a sensation of effort when trying to waggle our ears, or to move psychokinetically something outside our body? How far can we even try to do these things? We don't know where to begin. Bergson would say that anything we do experience on these occasions will result from certain muscles we actually do move, perhaps in trying to concentrate our attention on the task concerned. Furthermore the different degrees that we seem to experience of the sensation of effort he attributes to two things, a greater number of sensations in the general area concerned, and a qualitative change in some of them (TF, 26). These I will come back to, but first let me complete the catalogue of cases he considers.

Intermediate between the cases we have mentioned so far there are various cases of psychic states which are accompanied by muscular contractions and peripheral sensations where these are coordinated by an idea, either speculative, to give intellectual effort or attention, or practical, to give various emotions and desires. (In 1902 Bergson devoted a special study to intellectual effort, which he later reprinted in the book *Mind-Energy*.) Then finally there are the pure sensations, where the differences of intensity seem to be linked solely to the external cause.

In all these cases Bergson analyses apparent differences of intensity of the psychological elements by appealing to the numbers of such elements and to qualitative changes in them, or else to external causes as they are perceived or hypothesised by us. What he rejects is an appeal to the external causes themselves. This leads him on to end the first chapter with an attack on the 'psychophysics' of Fechner and Delboeuf, who tried to provide units in terms of which sensations themselves could be measured, and thus tried to reduce intensive magnitudes to extensive ones – a process Bergson thought as natural as it was illegitimate, the notion of an intensive magnitude being an unhappy halfway house between that of an extensive magnitude and that of what is not a magnitude at all. His own procedure is one of replacement rather than reduction.

3. Sensations

Let us take pains as perhaps the most obvious example. The claim is not that no pain can differ from another, but that pains are not magnitudes. If we have a mild pain and then a pain otherwise like it but more severe, how is this situation to be analysed? For analysis is what Bergson calls for. In a discussion of his views in 1901 he answered some objections raised by Elie Halévy and said, 'In *Time and Free Will* I criticised the notion of intensity in psychology not as false but as demanding to be interpreted. No-one can deny that a psychological state has an intensity' (M, 491).

What he is rejecting is intensive magnitudes, for any magnitude ought to be measurable. Now measuring 'implies that we really or ideally superpose two objects one on another a certain number of times' (CE, 230 – 1). As we saw earlier, we cannot superpose pains in this way, whether we experience them in succession or even simultaneously, say one in each arm. The concession that we need only superpose them 'ideally' will not help; for we cannot, even in imagination, subtract from one of them a bit which is 'equal' to the other and then consider some part of the first one which is left.

All we have so far is two sensations which differ in quality, as though we had an ache in one arm and an itch in the other. Sensations, however, do not arise simply as brute phenomena without connexions with anything else. They affect us, and we react to them in various ways; and they also represent their causes to us. Not all sensations need do both of these things. But Bergson seems to assume that all sensations, or all that vary in intensity, do at least one of them. He distinguishes between affective and representative sensations and seems to assume that the distinction is exhaustive, at least for cases where intensity comes in (TF, 32).

The distinction does not seem to be vital in itself, and Bergson allows a lot of overlap between its terms, but it gives us some variables to play with. Even without sensations our bodies react to stimuli, as we know from cases where we touch a very hot object and spring away from it before feeling any pain. But if we are to act freely, to be in charge of our actions and able to organise and direct them, it may be useful to us to know what these bodily reactions will be, or what they would be if we did nothing to control the course they take. So pleasure and pain may serve as signs of these reactions, both those already occurring and incipient ones (TF, 33 – 4). This typically pragmatist argument gives Bergson a way of accounting for at least some cases of different intensities among pains. For bodily movements, at any rate those of the kind relevant here, are things we are conscious of, in a way we are not conscious of the molecular movements that go to make up the stimulus. We can therefore avoid the objection that a physical phenomenon and a state of consciousness have nothing in common that could make changes of magnitude in the former appear to us as changes of magnitude in the latter. For bodily movements are conscious phenomena, as we said, and yet can involve magnitude in a relevant way. (See TF, 34 – 5, 32 – 3.) The way they do involve it is by the extent of the organism which they affect, and, what comes to much the same thing, by their number. When we have a severe pain we react in a complex set of ways, involving 'muscular contractions, organic movements of every kind' (TF, 35); 'we estimate the intensity of a pain by the larger or smaller part of the organism which takes interest in it' (ibid.). Quoting empirical evidence that 'the slighter the pain, the more precisely it is referred to a particular spot; if it becomes more intense, it is referred to the whole of the member affected' (TF, 36) Bergson proceeds to make this a matter of definition.

The other kind of sensation, the representative, can also be illustrated by a pain, but this time by one of whose cause we are fully aware. If we prick our hand with a pin more and more deeply we shall feel (Bergson assures us) first a tickling, then a touch, a prick, a pain localised at a point, and the spreading of this pain (TF, 42 – 3). Unreflectively we speak of the spreading of one and the same sensation, but reflection tells us that we are dealing with many qualitatively distinct sensations. The reason is that we localise in the sensation the progressive effort of the other hand which does the pricking, and so unconsciously interpret quality as quantity, intensity as magnitude.

One more case is worth mentioning, because it involves Bergson in appealing to something that elsewhere he vigorously criticises: the association of ideas. In dealing with intensity of sound he claims that, leaving aside certain vibratory effects, a sound's loudness is simply an indefinable quality, but that we interpret this as quantity because we have so often obtained such sounds by the expenditure of a certain quantity of effort. (See TF, 43 – 6. We have seen already how effort itself is supposed to have degrees.)

4. The issue over intensive magnitudes

How much of all this can we accept? Let us start with the last case, involving association. It is true that if we can make a noise by hitting a drum we can make a louder noise by hitting it harder. But could it really be because of this that we regard it as louder? Most sounds we hear we have no idea how to produce at all – as most pin-prickings are not self-inflicted. And if it is said that we transfer to other cases the differences in loudness that occur among sounds we can produce, notably vocal ones, which we can produce from birth, then how is it we know what to transfer? If non-vocal sounds vary in indefinable qualitative respects, what is the quality that we transfer from our own vocal screamings, and perhaps cot-bangings and rattle-shakings, so that we know which ones to call loud and which soft? The argument must be this: There is a certain generic property, which we can call sonority, that comes in various determinate forms, A, B, C, ... We notice that noises which we can make using only a small amount of effort have, say, quality E. Those which we make using a bit more effort have quality D, those which require still more effort have quality C, and so on. Regarding our own productions of these sounds as slightly effortful, fairly effortful, pretty effortful, and so on, we thereupon regard the sounds themselves as slightly loud, fairly loud, pretty loud, and so on. Then we notice these same qualities A, B, C, . . . , in sounds that we have not produced but could produce, and then finally in other sounds that we couldn't produce, but in all these cases we transfer the same terms 'slightly loud', 'fairly loud', etc., to them.

Now how are these determinate qualities, A, B, C, . . . , related among themselves intrinsically? Are they ordered, so that we can say intuitively that B is between A and C, but C not between A and B? Or are they simply different, like the tastes of fruit? For who could say that the taste of bananas is between that of apricots and that of cherries? But here we come to a dilemma. For, if the qualities are simply different, then we must presumably associate each quality with its appropriate degree of effort by a separate associative act, and it would be simply a brute fact that the association held in each case. This seems most implausible. For how, apart from anything else, could we extrapolate to degrees of loudness that we had never actually produced ourselves? But if the qualities are already ordered then why do we need the association with degrees of effort in the first place? Because it explains why we speak of the intensity of a sound as a magnitude, we are told (TF, 44). Since Bergson thinks the only genuine magnitudes are extensive and measurable, what we seem to have is a way of measuring the loudness, by measuring the degrees of effort. Now if we do talk of measuring loudness, and call one sound twice as loud as another, perhaps we do need some special explanation for how we come to do so. But the point of calling loudness an intensive magnitude is that it is not measurable though it is orderable. In fact is not Bergson admitting all that we really want him to admit, if he allows that loudness is orderable? Different cases of loudness are not superposable, but why need they be? Is he saying more than that intensive magnitudes are not extensive ones?

What is really meant by intensive magnitudes, and what is Bergson really trying to maintain? While discussing representative sensations he mentions flavours, something he says rather little about on the whole:

Between flavours which are more or less bitter you will hardly

> distinguish anything but differences of quality; they are like different shades of the same colour. But these differences of quality are at once interpreted as differences of quantity, because of their affective character and the more or less pronounced movements of reaction, pleasure or repugnance, which they suggest to us. (TF, 39)

The comparison with colour is puzzling. Are the 'shades' ('nuances') degrees of light and dark, or hues? The use of 'more or less' suggests the former. But then we have a feature in terms of which the colours can be ordered. How far can hues be ordered? As a whole they cannot be ordered along a dyadic scale, but only in a more complicated arrangement. But the hues of one colour can be ordered dyadically, and there is nothing to stop us doing the same with two colours which shade into each other, like blue and green. Can one then say that the hues between blue and green form an intensive magnitude? A changing light could be said to be 'becoming bluer', and something cerulean could be called more blue than something viridian, but hardly twice as blue. A possible difference one might see between this sort of case and dyads like light/dark, hot/cold, hard/soft, is that in these dyads one term tends to appear as the positive pole and the other as the negative. We normally talk of degrees of light, heat or hardness, and not of darkness, cold or softness. Green and blue on the other hand seem to be entirely symmetrical, so that if 'magnitude' means 'bigness' then it would seem equally proper to assign large values to the blue end and small to the green as vice versa, which creates a certain tension. But this seems to be rather a subjective matter – the sort of thing Bergson himself would be only too happy to explain in terms of associations derived from our scientific knowledge. The Greeks after all thought of hot and cold, at least, as equally positive terms. On this view then an intensive magnitude seems to be simply a dyadic scale. Whether colours, or tastes, etc., are intensive magnitudes depends on how we look at them. Taken as wholes they are not, but intensive magnitudes can be carved out of them to the extent that dyadic scales can be carved out.

But that is not all there is to the matter. If cerulean is bluer than viridian it is more blue. But 'more blue' is ambiguous. It might mean 'more of blue'; a delphinium is (contains, constitutes) more blue than a bluebell because it is bigger. Or it might mean somehow 'more thoroughly blue'. When Plato contrasted the extent and purity of pleasure by saying that a large white patch was no whiter than a small patch of the same shade he was preferring the second of these senses. (See *Philebus* 53b.) Bergson evidently prefers the first, and thinks that the second is only legitimate if it can somehow be replaced by the first. Summing up his earlier discussion towards the end of *Time and Free Will* he says

> [W]e found that psychic phenomena were in themselves pure quality or qualitative multiplicity, and that, on the other hand, their cause situated in space was quantity. In so far as this quality becomes the sign of the quantity and we suspect the presence of the latter behind the former, we call it intensity. The intensity of a simple state, therefore, is not quantity but its qualitative sign. (TF, 224)

(See also TF, 90.) To keep this consistent with the 'preliminary remarks' we must presumably distinguish the external cause in itself, which we need not know about, and the external cause as experienced by us, i.e. as a multiplicity of sensations, reactions, etc., or else as hypothesised by us. (See TF, 4 – 7, and p3 above.) But in what sense is the quantity 'behind' the quality? To see this we must go back to some of the other things Bergson said about affective and representative sensations. But before doing so let me clear away a possible source of confusion. Someone might object to calling cerulean more blue than viridian because viridian is a kind of green, albeit a bluish green, and so is not blue at all, not even slightly blue. But this would be like saying water a few degrees above freezing is not warm, not even slightly warm, or that an event with a probability of twenty per cent is not probable, not even slightly probable, if 'slightly probable' means 'slightly more probable than not'. There is a systematic ambiguity in such terms, which should not cause any real difficulty.

What we saw in the two pain examples, the affective and the representative, was an appeal to multiplicity. We had references to 'muscular contractions, organic movements of every kind' (TF, 35) and to the extent of the organism that 'took interest' in the pain, and then, in the case of the pin-pricking, to the spreading of a pain as involving many qualitatively distinct sensations. Many sensations are more than one. They can be distinguished, either by the place in the body where they occur or by being qualitatively different, and then they form a magnitude, which in principle could be measured, if one is willing to treat the things one counts as homogeneous units for the sake of the counting. But the pain itself has no countable parts each of which is painful, and so it is not a measurable magnitude. We can measure its severity only if we replace this by something else, an extensive plurality of sensations of muscular movement etc. This, as I interpret him, is what Bergson means by saying, as I quoted him earlier, that a psychological state does have intensity but is not a magnitude. It has intensity because this replacement can occur.

5. Psychophysics

But perhaps this is premature. Perhaps the sensation can after all be measured, by being shown to have units of its own. Bergson considers two attempts to do this, by the methods of psychophysics, associated with Delboeuf and Fechner, and using the example of the intensity of light. (See TF, 52ff.) The general idea is to use either intervals between shades of grey that are judged subjectively to be equal (Delboeuf) or minimum perceptible differences between such shades (Fechner). For Bergson the problem is basically to show that 'two sensations can be equal without being identical' (TF, 57). (Cf. TF, 63.) He does not deny that we can consider a series of differences between sensations and construct measures for the sensations by adding them together; but though we *can* do this, it does not give us anything but an illusory measure because it is arbitrary. We can construct a definition of a measure in this way, but, as he quotes Jules Tannery as saying, this 'is as legitimate as it is arbitrary' (TF, 67). We can count the shades between two colours, but what will guarantee that the successive differences are magnitudes, let alone equal magnitudes? (See TF, 58.) Two sensations S and S' may be different, but how can we form the interval between them, as we could between two numbers by making the subtraction, S'-S (TF, 66)?

This is all very well, we might say, but isn't Bergson (and Tannery too) asking too much? How can any system of measurement operate except by taking units, and must not the units ultimately be arbitrary, not simply as regards their size but as regards their equality? Units *qua* units must be indivisible and, as for the arbitrariness of the assumption that they are equal, how could we measure, say, time except by assuming the equality of, say, successive sidereal days? So far Bergson would be unimpressed. Who says that we *can* legitimately measure time? No doubt units *qua* units must be indivisible, but we may only be entitled to take them as so after we have established that they are indeed suitable units. Inches, *qua* units of measurement, are indivisible. In taking them as our units we ignore finer subdivisions of our ruler. But if two allegedly inch-long pieces of wood are superposed and one overlaps the other they cannot both be used as units. What similar tests could we apply to sensations? '[T]wo different sensations cannot be said to be equal unless some identical residuum remains after the elimination of their qualitative difference' (TF, 64). But what could such a residuum be? And, we might add, if all we know about two sensations is that they are different, and we cannot get from one to the other by adding to it (TF, 66), then why should the difference between two shades of colour be more suitable to use as a unit than the difference between, say, a colour and a sound?

For Fechner there is a common residuum, not to two sensations but to two differences between pairs of sensations, namely the property of being minimum differences (TF, 64). As we saw two paragraphs ago Bergson questioned whether these differences were magnitudes. He might also have asked why the difference between a colour and a sound should not also count as a 'minimum' difference, since one can hardly insert any further terms between them – yet to call this a minimum difference is hardly in the spirit of what we want. But there are two points that an opponent of Bergson might make here. First, the fact that two differences are minimum differences does not prove, or indeed give any reason for thinking, that they are equal, but it might give a reason for treating them as equal; it might show that it was not arbitrary to do so. Indeed one might go further and suggest that Bergson's own appeal to the role of superposition and identity in measuring extensive magnitudes involves assuming that our rulers keep their length when moved around in space. Furthermore the whole notion of superposition is problematic if we take seriously an argument he several times appeals to later that even distances that seem small to us would seem large to a microbe. (See e.g. M, 109 – 10, in *Durée et Simultanéité*.) However, he might answer that it would be arbitrary to assume without reason that rulers changed their length while other things didn't, and that superposition of rulers was a different thing from the bringing together of clocks and objects that he was concerned with in *Durée et Simultanéité*: clocks will always keep a distance between their centres, where the mechanism is, but with rulers and ruled the operative parts are the edges, which *can* coincide. Let us not ask further how far this is true.

The second point is that there is an intuitive resemblance between Delboeuf's shades of grey, or between neighbouring colours, which there is not between a colour and a sound; and this again might serve to make Delboeuf's units non-arbitrary ones. The choice of the sidereal day for measuring time is not arbitrary, because it fits together with a whole lot of other repetitions, notably those involved in our own metabolism, so that sidereal days strike us intuitively as equal. They need not be perfect in this respect (intuitively the solar day would be better), and as Waismann pointed out we could have found it useful to take instead the periodicity of thunderstorms (Waismann 1949 – 53). But the sidereal day, or the quartz clock for that matter, is near enough to the intuitively preferable solar day to allow its scientific convenience to prevail.

Bergson in fact reduces Delboeuf's use of intuitive resemblance to Fechner's use of minima, because in assigning equality to two differences we are implicitly inserting the same number of minimum steps into each of them (TF, 59). It is not obvious that we need do this. If I say that blue is nearer to green than to red, need I have in mind, even implicitly, the number of shades between blue and green and between blue and red? And suppose I say, as I surely could with equal plausibility, that green is nearer to blue than to red: there are no shades between green and red by a direct route. There are no reddish greens or greenish reds, unless one insists that viridian is a reddish green just because it is a bluish green and one can get from green to red via blue – but then one would have to say that if leaf-green is a yellowish green it too is a reddish green, since one can get from green to red via yellow. Should one say instead that resemblance is partial identity, the identity here being in respect of properties: blue and green are both cool colours? But what property makes yellow nearer to orange than to crimson? Resemblance at least in these cases seems to be unanalysable. Bergson points out that we do not in fact generate magnitudes from hues as we generate the light/dark scale from shades of grey because we experience greys along with experiencing systematic variation in degrees of illumination (TF, 58). Presumably he would add that we do not similarly associate hues with varying frequencies.

6. Conclusions

Should we conclude from the last two paragraphs that sensations are not only magnitudes but extensive magnitudes? Bergson considered that if we took the first step it would be hard consistently to avoid taking the second. (See TF, 70 – 2). Once you grant that sensations are magnitudes enough to admit of degrees it is hard not to go further and try to measure them, as the psychophysicists do. But, though it may be natural to attempt this, we do not have to. We can *call* one sensation twice as loud or hot or even blue as another if it is separated by twice as many minimal perceptible steps from some state deemed to involve an absence of loudness or heat or blueness, but there is not the same point in doing so as with things like length. Not only are the units more variable and not checkable by superposition, but the magnitudes are not obviously addible. If I engineer a sensation of heat in myself I cannot then add another similar sensation to produce one twice as hot. I can engineer a sensation twice as many units removed from the zero state as the first sensation, but this is not adding sensations. The same applies to blue. In the case of loudness, if I listen to an oboe and then add in another oboe the resulting sensation will no doubt be louder, but it need not be twice as many units from silence. The sensations might interfere with each other.

But why need we assume from all this that the sensations cannot be ordered, and so form an intensive magnitude? Bergson, as we have seen, thinks they can only be ordered parasitically, because some other magnitudes can be ordered and also measured, such as some physical feature of the cause that we are conscious of or infer, or the number of muscles etc. involved in our actual or incipient reactions. He does not, I think, mention hardness as a candidate for being an intensive magnitude. Surely things can be arranged in a 'scratching order', provided we do not find anomalies such as a trio where A will scratch B but not C, B will scratch C but not A, and C will scratch A but not B. But can we guarantee that between two apparently adjacent terms a third term will not at some time appear, so that we shall have to keep revising our units? We might avoid this by making our unit a certain depth of scratch got by applying a certain force for a certain time, so that A is twice as hard as B if A scratches C twice as deep in a given time as B does. But hardness is not additive: we cannot add hardnesses together. This suggests that the sense in which A can be twice as hard as B is rather a secondary or derivative sense and we have not really got an extensive magnitude here. But why use this fact, if it is one, to stop us calling hardness a magnitude at all?

If we deny the existence of intensive magnitudes we face the question why there seem to be such things, and how there can even seem to be such things if the notion itself is senseless. In his discussion of joy Bergson points to the varying numbers of our psychic states which it pervades, and says we

> thus set up points of division in the interval which separates two successive forms of joy, and this gradual transition from one to the other makes them appear in their turn as different intensities of one and the same thing, which is thus supposed to change in magnitude. (TF, 11)

But why should a gradual transition from one thing to another make them seem to have a property it is senseless for them to have? It doesn't seem to help when he makes the same point about aesthetic feelings and adds, 'this qualitative progress . . . we interpret as a change of magnitude, because we like simple thoughts and because our language is ill-suited to render the subtleties of psychological analysis' (TF, 13). All this is surely much too Humean.

Bergson has warned us that we ought not to assume too easily that we have a difference in magnitude of a sensation when in fact we have only a difference in quality. But this would be different from assuming we have a magnitude when such a notion is senseless. Let us return to pains, which we left by the wayside when discussing loudness. Bergson tried to analyse the intensity of pains in terms of the extent of the body affected or the number of our actual or incipient reactions. But, though a severe pain may spread in the way he suggests, it simply is not true that we can't have two pains differing in severity but not differing in extent, and we can have sharp localised pains and diffuse aches. Nor is it true that our reactions must always be proportional to the pains' 'severity', at any rate if both pains are not too severe. It would be mere dogmatism to insist that there *must* be incipient or unconscious reactions of a suitable kind in each case, and what reactions there are are often plainly not confined to the area where the pain is; even if a beating heart is part of one's fear, a visit to the doctor is not part of one's stomach-ache, and neither is the instinctive soothing movement of one's hand. A tensing of the muscles may be part of the total experience but is not part of the pain, since one can often untense them without lessening it. Also this view destroys the idea that we react *to* the pain and *to* its severity. Bergson explicitly refuses to go all the way with James and say 'that the emotion of rage is reducible to the sum of these organic sensations' (TF, 29), but he adds that 'the growing intensity of the state itself is, we believe, nothing but the deeper and deeper disturbance of the organism' (ibid.). Talking of pity he says: 'The increasing intensity of pity thus consists in a qualitative progress, in a transition from repugnance to fear, from fear to sympathy, and from sympathy itself to humility' (TF, 19). He thinks the last of these stages represents 'true' pity, which it may – but what has this to do with 'increasing intensity'? The progress is surely from not really pitying someone to pitying them in the proper sense, not from pitying them a little to pitying them a lot.

Bergson, as I say, has warned us against a confusion that we might indeed sometimes make. But I suggest he has unduly ignored the intuitive resemblances that strike us between qualities and enable us to arrange them in orders.

Finally let me draw together some of the conclusions of the last few pages. Bergson could be thought to have shown only that intensive magnitudes are not extensive ones. I have suggested his position is rather this: he shows that loudness is not intrinsically measurable; it is not an extensive magnitude, and is measurable only by being associated with something like effort. But he seems to think he has shown that it is not intrinsically orderable, and is therefore not an intensive magnitude, though it does have intensity because it can be non-intrinsically ordered by reference to effort etc. Similarly 'more intensely blue', he thinks, can only mean something like 'more of blue', and because it can mean this we again have intensity, but not intensive magnitude. One might think blue could be measurable; but he rejects Delboeuf's intuitive mid-points because they seem to reduce to Fechner's minimum differences, which themselves cannot provide units without arbitrariness. I have suggested that what matters here is not that they are arbitrary but that they cannot be added or manipulated. But blues can still be intrinsically orderable, and so can form an intensive magnitude; and if the notion of intensive magnitude is incoherent why, and indeed how, has it had such popularity?

I said, however, at the beginning that this first chapter contains what could be called the germ of the system. What underlies the attack on intensive magnitudes is the conviction that magnitude is tied to space. Bergson says of a view to be rejected that it is held 'as though it were permissible still to speak of magnitude where there is neither multiplicity nor space!' (TF, 9) and he plainly implies that magnitude involves space (TF, 21). The second chapter of *Time and Free Will* takes us straight on to Bergson's fundamental positions on space and time.

II

Space and time

1. Introduction

It is quite common in philosophy to treat space and time in unison. We ask whether space and time are real, whether they are absolute or relative, whether they are finite or infinite, and we compare various sorts of things according as they do or do not 'exist in space and time'; in each of these cases we tend to assume that the same answer must be given for both space and time. Of course we recognise some differences. Space has three dimensions, time only one, while time has a built-in direction and space has not. Kant regarded space as the form of outer sense and time as the form of inner sense. Space seems to be correlated with objects in a way that time is correlated with events; and not everyone would agree with R. Taylor (1955) that things can move around in time just as they can in space. But there is a deep strand in our thinking that wants to assimilate space and time at least as far as possible, a strand perhaps only strengthened by the notion of space-time that has emerged in modern physics.

All this is quite alien to Bergson. He does indeed make certain comparisons between space and time, notably in seeing two forms of each of them. But the comparison consists in making them approach each other from two quite opposite points of view: there is a sort of second-best form of time which is rather like space and a second-best form of space which is rather like time. Basically space and time play opposite roles in his philosophy, which might even be regarded as a dualism of space and time.

2. Space and counting

Time is the term most prominently associated with Bergson's philosophy, but let us start by discussing space. Bergson closely associates space with homogeneity and with plurality, and with the extensive as against the intensive. He seems to have been impressed with a certain paradox of counting, which we have already hinted at: how can we count things unless they are there together to be counted? And how can we count things, or talk of a plurality of things, unless they are the same things – as we might put it in modern terms, how can we count things unless we subsume them under a single concept? And finally if they are there together and are the same how are they to be distinguished from each other? The answer is provided by space, as a sort of medium of pluralisation.

This paradox may not seem very compelling; in fact its force may not seem very clear, until we see the use Bergson makes of it, which is to argue that *all* genuine plurality involves space. We normally think of time as admitting of measurable plurality. In fact it is regarded, along with mass and length, as one of the three basic extensive magnitudes. But Bergson will not allow us to count and add units of time without further ado, for no two units of time can exist together for us to operate on them. In order to measure time we must first spatialise it, i.e. treat it as though it were space. More generally this applies to number, and Bergson begins his second chapter by discussing number, which he defines as 'a collection of units' or 'the synthesis of the one and the many' (TF, 75). Russell (1914) accuses him of confusing number, numbers, and the numbered, and the charge is fair enough, in that he speaks of three as a collection of three units without asking such questions as: Which three units? He confuses mathematical addition with actual juxtaposition. But I doubt if the confusion affects his main concern, for he could claim that we cannot have the concept of three without having that of a threefold group.

But merely calling number a collection of units is not enough, for the units must be, or be treated as, identical with each other and yet also distinct, or they would merge into a single unit (TF, 76 – 7). Here we are back with our paradox, and to solve it we must bring in space, for if we are to count, say, fifty sheep we must either 'include them all in the same image, and it follows as a necessary consequence that we place them side by side in an ideal space, or else we repeat fifty times in succession the image of a single one' (TF, 77) – but this in fact comes to the same thing because we must retain the successive images and put them alongside each other or we will just repeat the number one.

But perhaps all this is just a feature of the example. Obviously sheep must exist in space. Suppose, however, the sheep are all of different colours and we count them accordingly: 'The black one, the white one, . . . ', in effect counting colours. Why do we now need space? Bergson, I think, would say two things. First, we are indeed counting colours, if anything, and not sheep. We may be enumerating sheep, but will not arrive at a total, as one enumerates soldiers, but does not count them, in calling the roll (TF, 76). And second, counting colours, or any other abstract things, still involves treating them as in space. Otherwise we shall again be enumerating rather than counting. We shall have fallen into the habit of 'counting in time rather than in space' (TF, 78). But if the things in question, whether colours or numbers or anything else, are actually to be counted they must remain together while the counting goes on, and this involves space: 'We involuntarily fix at a point in space each of the moments which we count, and it is only on this condition that the abstract units come to form a sum' (TF, 79). Of course the moments themselves are fleeting, but 'we are not dealing with these moments themselves, since they have vanished for ever, but with the lasting traces which they seem to have left in space on their passage through it' (ibid.), i.e. with a mental image. In fact he adds that, 'every clear idea of number implies a visual image in space' (ibid.) One might wonder why the image has to be visual ('une vision') – surely blind people can count – but that is a minor point.

Slightly more important is his admission that we generally dispense with this image after using it for the first two or three numbers, because we know we always could use it to represent the others if we needed to. Evidently then the actual act of counting does not itself involve images, though it presupposes their possibility. Presumably what it presupposes is that the things counted are spatial, from which their imageability follows.

There is an ambiguity here, however. No doubt if any given countable element is spatial we can image it. But it does not follow that we can image the whole sets of things we are counting. Try imaging 600, to go no further. Perhaps we start by imaging ten dots in a row and then add nine other such rows to form a block. Then we image six of these blocks – but do we still image each block as containing its ten rows of ten dots each? Obviously the difficulty is worse for larger numbers, and it is not optional that we dispense with the images. Yet we can understand such numbers perfectly well. The most Bergson could argue is that every clear idea of number implies *some* image or images in space – perhaps the same ones for each number, namely those that are needed to get counting going in the first place.

But, if we *can* count without imaging, what are we doing when we do, and why does Bergson say we could call upon the image 'if we needed it' (TF, 79)? When might we need it? It seems that he is making a logical rather than a psychological point here. Countables must exist together, whatever the actual procedure of counting might consist in. This is one reason he has for making the countables spatial. But he does not always distinguish it clearly from another. In the footnote that opens this second chapter he considers a possible case where we seem not to need space, namely when we count simultaneous impressions received by several senses. But he says of this, 'We either leave these sensations their specific differences, which amounts to saying that we do not count them; or else we eliminate their differences, and then how are we to distinguish them if not by their position or that of their symbols?'

Two points here call for comment. First, the reason given for needing space is not now that the things must coexist but that they must be individuatable, as though they were as like as two peas. But they weren't. They were radically different, being impressions of different senses, which *we* then assimilated by eliminating their differences. Bergson seems to assume that when we consider two different things as though they were the same we substitute for them two things that *are* the same. The second point arises from the last few words: 'or that of their symbols'. If the countables themselves need to be individuated we can hardly use their symbols for this purpose, for until they have been individuated how do we apply the symbols to them? And if it is indeed the symbols that must be distinguished in space then why need the countables themselves be distinguishable spatially (as opposed to some other way) as well? The point is of some importance because Bergson insists that the countables, to be counted rather than merely enumerated, must exist together, and denies for this reason that time is countable without further ado, as we saw. But if only the symbols need to exist together, which would presumably be why they had to be in space, then why should not time be countable after all? Bergson, it is true, thinks that we can count time or numbers in so far as we treat them as spatial, and it might be argued that this reference to spatialising the symbols is just what is meant by treating time or numbers as spatial. But it is hard to avoid thinking that he is not entirely clear about when he is talking of features that must belong to what is to be counted and when of features of the symbols we use in doing the counting. Suppose we allow that to count units of time we must have symbols of them together before our mind: does it follow that time itself, in all its evanescence, cannot be what we are counting? But, if it can, why should time not be an extensive magnitude? (On space and individuation cf. also Stewart 1911: 212 – 23.)

However, this comment arose from what might be a casual aside on Bergson's part, the words 'or that of their symbols'. He plainly does not in fact accept that time could be, in itself, an extensive magnitude, and seems in fact to think of the countable itself as what must be accessible all at once, whether this is the original thing we would like to count or something else we substitute for it.

This idea of substitution is a difficult one, and smacks of being used as a get-out. Surely either one can measure or one cannot; what is the point of trying to measure something else instead? Bergson admits that when we do 'abstract' time in this way we are 'really giving up time' (TF, 98). The procedure reminds one of how St Augustine tried to deal with the same problem of measuring what was not there to be measured by setting up a representation of the time in his own mind and measuring that instead, a procedure subject to the same objection of *ignoratio elenchi*: the wrong thing is being measured. With Bergson, however, it would be unfair to dismiss him by taking this point in isolation from the rest of his philosophy, which goes far beyond that of Augustine. In particular it would be unfair to condemn him for substituting something purely mental for something purely physical without asking how far he is willing to accept that the mental and physical form two separate realms at all. But the use of substitution remains problematic.

Let us sum up this discussion of counting. For Bergson countable things must exist together and so simultaneously. They must be the same, to fall under one concept and not be merely enumerated, yet they must be different, to form a plurality. Therefore they must exist in space – or if they do not then we must substitute for them other things that do. All this involves two assumptions: first, that the only way of being simultaneously together is to be in space, and, second, that to count things we must think of them as the same in some respect and ignore their qualitative differences – only to find that then we must bring in something to separate and individuate them. The first assumption ignores the admittedly controversial possibility that sounds might form a non-spatial plurality. (Cf. TF, 86 – 7.) Later Bergson insists that 'all sensation is extensive in a certain measure' (CE, 213). (Cf. Čapek 1971: 319, who refers to this as p.221 of the paperback.) But to bring in this point here would be to analyse the spatial in terms of auditory plurality rather than vice versa. The second assumption amounts to this, that in abstracting from the differences between things one thereby creates a new set of things, which need to be differentiated in turn. But there seems no reason to make this assumption, and the problem would merely recur, with spatial properties instead of the original differences.

3. Space and extension. Kinds of multiplicity

Space for Bergson is associated with homogeneity and time with heterogeneity. In space things exist separate from and alongside each other; in time they interpenetrate and are never completely independent. An experience, for instance, is always influenced by all previous experiences. But he also insists that 'every homogeneous and unbounded medium will be space' (TF, 98), because two such mediums could not be distinguished from each other. This is because homogeneity in this case means the absence of every quality. This tempts one to think that the equation of space with the homogeneous is a matter of definition rather than of significant assertion, and indeed Bergson introduces it as such at TF, 98, though he immediately goes on to treat it as synthetic, not only by offering a reason for it (the one just mentioned) but by treating it as a matter for discussion whether time forms another such homogeneous medium. The issue is whether, given that we have a reason for postulating exactly one homogeneous medium, we explain what space is by identifying it as that medium or specify the medium by identifying it with an already familiar medium of space. Bergson seems more inclined to do the latter.

But the situation is complicated because he does not entirely and unambiguously identify space and extension. As I said earlier, he sees two kinds of space, albeit one is more thoroughly 'spatial' than the other. This distinction comes more in his later works. (See Čapek 1971: 208 – 10.) But already in *Time and Free Will* we read that 'We must thus distinguish between the perception of extensity [l'étendue] and the conception of space' (TF, 96). (The French has 'Il faudrait', not 'Il faut', but Bergson is clearly giving his own view. Cf. also MM, 245.) Bergson thinks here of sensations in themselves as purely qualitative and not partaking in extension, though later he insists that they do all partake in extension. (See e.g. MM, 288.) In *Time and Free Will* it is the activity of the mind which 'perceives under the form of extensive homogeneity what is given it as qualitative heterogeneity' (TF, 95). His example is of the difference between the impressions made on our retina by two points of a homogeneous surface. He thinks that this difference is in itself basically a qualitative one, but adds that 'there must be within the qualities themselves which differentiate two sensations some reason why they occupy this or that definite position in space' (TF, 96).

Perception *qua* something active involves extensity, but space is more conceptual or intellectual than extensity, and Bergson goes on to indulge in a remarkable hypothesis that animals perceive the world in a less purely spatial way than we do, and that the reason they are often so good at navigating in conditions that would defeat humans is that the different spatial directions have more of a qualitative content for them, so that space is less homogeneous for them than for us! No doubt north has a peculiar 'feel' for a magnetically sensitive animal, as down does for ourselves, but one would like to know whether north would keep this 'feel' as the magnetic pole wanders around the globe, and indeed just what Bergson means by 'directions': surely not those of absolute space? But this point about animals is well in keeping with the distinction Bergson will eventually draw between intuition and instinct on the one hand and intelligence on the other, the first pair being associated with duration and intelligence with space.

The distinction between space and extension cannot be fully grasped in isolation and without reference to what Bergson says about time and motion, on which in turn it throws some light. Extension or extensity he usually calls 'l'étendue', though his terminology is not always very consistent (and not always faithfully reflected by the translators of *Matter and Memory*). It is more something in the world, as its link with perception would suggest, though to say this is not fully adequate: 'That which is given, that which is real, is something intermediate between divided extension and pure inextension. It is what we have termed the *extensive* ['l'*extensif*']' (MM, 326). What Bergson wants to avoid, at any rate in his later works, is a sharp opposition between an extension really divided into corpuscles etc. and a consciousness whose sensations are in themselves inextensive but project themselves into space. Our understanding may artificially represent things to us like this; Bergson seems to have in mind the way Berkeley refused to allow the data of sight to have extension, at least in the third dimension (MM, 286, 289), while realists denied it to any sensations (MM, 283 – 4). But if things really were so the correspondence between the two realms would be inexplicable. (See MM, 326, and cf. MM, 282 – 8.) There is no 'abrupt transition from that which is purely extended to that which is not extended at all' (MM, 284). (Cf. MM, 293 – 4.) The sort of halfway house Bergson is looking for becomes a bit clearer when he talks of the need to obviate 'a confusion of concrete and indivisible extensity with the divisible space which underlies it' (MM, 293 – 4). Anything which is to be real must be undivided and possess the unity of a whole. Yet space is essentially the divisible and provides the framework for counting, which is homogeneous in the sense that countables must be regarded as exactly similar to each other for the purpose of being counted. Since real things are not exactly similar in this way and are not 'cut off from each other with a hatchet', as Anaxagoras put it in a somewhat different context, spatiality must be something of an ideal limit and cannot wholeheartedly apply to real things. (See fragment 8 of Anaxagoras.)

This at least is the view of *Matter and Memory*. In *Time and Free Will* the world in itself is more definitely spatial and is devoid of real temporality: 'Thus, within our ego, there is succession without mutual externality; outside the ego, in pure space, mutual externality without succession' (TF, 108). Time and space go together in both works. The more spatiality the world has, the less temporality it has, and vice versa. If the outer world is completely spatial, it is completely non-temporal, which seems odd. But what Bergson means is not that there are no different states of the world at different times, for '[t]he sounds of [a] bell certainly reach me one after the other' (TF, 86). But there is no *transition* from one state to another. They are simply laid out side by side, entirely external to each other and therefore unrelated. In *Matter and Memory* this view is moderated, and taken to refer not so much to the world in itself as to the way in which intelligence inevitably is led to regard the world, which it does in order to act on the world, as we will see later.

But this double view of space, and the associated double view of time, which we shall come to, go along with a double view of multiplicity. (See TF, 85 – 7, 121, 128 – 129.) There is a multiplicity of material objects and a multiplicity of states of consciousness. The first of these is ordinary multiplicity, but the second is harder to grasp, and seems to be more like qualitative variegatedness. It is often called qualitative multiplicity. One thing Bergson seems to be referring to is the richness of an experience which has many different features, but where not only the experience as a whole but the features as well would not be what they are without each other. An ordinary multiplicity is of objects which are severally self-sufficient, and he wants to contrast this with the notion of a complex but indissoluble whole. But one may well doubt whether the distinction as thus portrayed is really viable. Take an experience, say, of seeing a colour. The experience will involve hue, tone, saturation, shape, size, and location, and will probably be compresent with other visual experiences, as well as experiences of the other senses, and of memory, thought, etc., all of which may contribute to making the original experience what it is. None of these features or experiences could exist unchanged on its own, we may suppose. But have we really two kinds of multiplicity here? The features and accompaniments can be counted under just those headings, 'features' and 'accompaniments'. This involves a certain amount of abstracting, as Bergson would be the first to agree, since what is being counted could not exist in isolation. But this does not affect the counting process itself, and 'multiplicity' is a term categorially linked to counting: where we have a multiplicity we can at least start counting. Even real numbers we can count to the extent that we can pick them out. To point to a difference among multiplicities between those whose elements can exist in isolation and those whose elements are mere abstractions is one thing. But the difference is really one between the elements, not between kinds of multiplicity. If a qualitative multiplicity just is one whose elements are qualities, that is fair enough, but it is no more a special kind of multiplicity than a multiplicity of dogs constitutes a special 'canine multiplicity'.

But to say this is not to say that Bergson has no point. I have assumed our dogs are quite independent of each other. Now take two brothers, each of whom would have been different in character had he been an only child, or a husband and wife, each of whom only is such because of the other one. Here we have cases of causal and logical dependence, but the two elements in each case clearly form a multiplicity. They can, however, also be regarded as a single entity, a family, or a married couple, just as the dogs can be regarded as a pair. But the grouping of the dogs is in a sense more abstract or artificial than the other groupings. There is a purely external relation of similarity between them. We have more occasion to talk of the brothers as a family, or of the married people as a couple, than of the dogs as a pair, even if this is only a matter of degree. Now take one's experiences during a day, with all their complex interlockings. It may be more arbitrary how we divide them up, but if we do divide them up we have a multiplicity. Alternatively we may talk of a single stream of experiences which has a multiplicity of features to it. We may talk of a multiplicity of experiences or of one stream (or slice) of experience with a multiplicity of features. In other words we have a choice about what we posit a multiplicity of, and there may be good reasons for making one choice rather than the other, for emphasising, perhaps, the unity and connectedness underlying our experiences and so for talking of one stream of experience. But a multiplicity is a multiplicity in each case. The same holds if there is no non-arbitrary way of selecting what to count. Take a coloured expanse varying smoothly in hue from one end to the other, or an experience similarly varying over time. It may be arbitrary how many colours we distinguish but we can still distinguish and count them, and even if they are non-denumerable like the points on a line we still presumably have a multiplicity; we certainly call them points (plural). The variegated experience can only be called a multiplicity in so far as we do consent to single out stages of it and count them, but Bergson could be justified in pointing to the arbitrariness of how we do this.

4. Duration

As I said earlier, Bergson's philosophy is predominantly a philosophy of time, or rather of duration ('durée'), as he usually calls it. In a letter of 1915 he writes:

> In my opinion every summary of my views will distort their general nature and will by doing so expose them to a host of objections, if it does not set out from in the first place, and constantly return to, what I regard as the core of the doctrine: the intuition of duration. The representation of a multiplicity in the form of 'reciprocal penetration', quite different from numerical multiplicity – the representation of a heterogeneous, qualitative and creative duration – is the point from which I set out and to which I have constantly returned.' (M, 1148)

What then is so special about duration? The main idea seems to be that it is not, like time as generally conceived, a set of discrete moments. It is not a set of anything at all in fact but rather a unified process. It has not the plurality of a set, which must be a set of things all alike in some respect and so homogeneous. It is heterogeneous, which presumably means that one bit of it is not the same as another bit. But what does 'same' mean here? One moment in ordinary time or one point in space is not numerically the same as another, but it is qualitatively the same in so far as they are both moments or points. Duration does not consist of things that are qualitatively the same in that sense. It follows that no state of anything that partakes of duration can ever be repeated – at least this follows if we interpret in a strong sense, as I think Bergson intends us to, the condition that duration does not consist of similar parts: not only does it not consist entirely of such parts but it does not contain any two such parts.

One of Bergson's favourite images to illustrate the nature of duration is that of a melody. In a lecture given in Oxford in 1911 he insists that though a melody takes time it is indivisible: 'if the melody stopped sooner it would no longer be the same sonorous whole, it would be another, equally indivisible' (CM, 174 (147)). He is evidently denying here that the melody could be abbreviated. Whether it could be accelerated is another question. (See §9 below.) A melody must consist of a plurality of notes, but Bergson is surely right in saying that equally the notes must be united into a single whole. They form what G.E. Moore would call an organic unity. The experience of hearing a melody will not be what it is – the experience of hearing that melody – unless it is taken all together. In this way it differs from something like a person, which again is something which endures and changes through time. A person whose life was exactly the same from moment to moment, who never developed or changed in any way, would hardly be a person at all. But a person is what he is at a given moment all the same. If I meet Smith at noon I meet the whole of Smith, or Smith as a whole, not a time-slice of a four-dimensional space-time worm – that would be Smith's life-history, not Smith. But if a car with its radio on whisks past me at noon I do not hear the melody as a whole, even if I hear enough of it to know what it was and to reconstruct it for myself. The melody in fact is rather more like the space-time worm, though I do not know if Bergson would welcome such a static-sounding comparison. The melody does not itself change, but it essentially involves change. Bergson would say it *is* change.

Just as there were two kinds of space, so there are two kinds of time for Bergson, heterogeneous duration and homogeneous time. Officially he separates 'duration' and 'time' in this way, e.g. in his 'Introduction to metaphysics' (CM, 211 – 12 (179 – 80)). But often in practice he uses 'duration' and 'time' indifferently for either notion. He does not deny that time involves plurality. The notes of the melody do indeed come to me at different times. Nevertheless time as so conceived becomes a sort of second-class time, which he calls 'imaginary', 'a fiction', and 'an idol of language' (MM, 274). In order to think about the world, still more to talk about it, as opposed to simply experiencing it, we have to 'freeze' it, i.e. apply concepts to it which can be applied repeatedly, whence they can be called universals. If language is to serve its purpose in fixing our thought and communicating it to others it must abstract and words must denote the same things for different speakers and for the same speaker at different times, and some words at least must be applicable to more than one thing. Thought and language involve generality, as Bergson says in the closing pages of the second chapter of *Time and Free Will*, though I have developed the argument more explicitly. As space and plurality in general involve homogeneous elements, so now does time as our experience over time gets broken up into bits that we can manipulate in thought and language.

This idea raises various problems. First, how does this distinction relate to that between the concrete and the abstract? The obvious answer is that real duration is concrete – Bergson sometimes refers to it as 'concrete time' or 'concrete duration' (e.g. TF, 214 – 20) – while homogeneous time is abstract; it is certainly what results when we abstract from our experience and consider things intellectually. Yet duration itself seems to be an abstract limit. At any rate in opening the third chapter of the much later *Durée et Simultanéité* with a summary of his views in this area as expounded in his earlier books he says of duration:

> A melody which we listen to with our eyes closed, thinking of nothing but of it, comes close to coinciding with this time which is the very fluidity of our internal life; but it still has too many qualities, too much determination, and first we would have to erase the difference between the sounds, and then abolish the distinctive characters of sound itself, keeping only the continuation of what precedes in what follows and the uninterrupted transition, multiplicity without divisibility and succession without separation, to recover in the end fundamental time. Such is immediately perceived duration, without which we would have no idea of time. (M, 98)

Here Bergson seems to be trying to get at an abstract idea of duration in a way not obviously consistent with his own philosophy. No doubt there is a difference between duration itself and an abstract definition of duration. But the above account suggests that duration itself is something abstract. Perhaps, in Bergson's philosophy, duration is something we cannot have an abstract idea of. Perhaps we can only experience it and not, or not adequately, talk about it. (Cf. M, 355.) To say this would be close enough to the spirit of Bergson's philosophy, since if philosophising is an intellectual activity, and therefore involves the 'freezing' of what it is talking about, it would not be surprising if difficulties arose when we tried to talk about the inherently unfrozen. Bergson's would not be the first philosophy to find difficulty in accounting for philosophising itself. But all this is to anticipate the distinction between intellect and intuition, which will come up later.

A more important, though not unconnected, problem concerns how duration is related to the world. We tend to think of time as that in which change occurs, and that which can be measured by changes, even if we also think that there could be no time without change – unlike Newton, for whom absolute time 'flows equably', whether anything is happening in it or not. (See Newton 1964: 81.) Duration for Bergson is certainly not an external framework of this kind. It cannot be measured, for reasons we have already discussed; and it must be distinguished from that 'homogeneous and impersonal duration, the same for everything and for every one, which flows onward, indifferent and void, external to all that endures' (MM, 274) which is what he calls an idol of language (ibid.) Bergson often insists that duration implies consciousness. Its 'essence is to flow without ceasing, and consequently not to exist except for a consciousness and a memory', he writes in a review in 1891 (M, 353); the 'consequently' prevents any confusion with Newton. Various writers see a divergence between *Time and Free Will* and Bergson's later books, where he modifies the extremeness of his initial dualism. (See Berthelot 1913: 248 – 50, Čapek 1971: 189 – 92, 208 – 10.) The issue concerns whether duration occurs in the physical world or is confined to the sphere of consciousness. *Time and Free Will* is unambiguous: 'Thus, within our ego, there is succession without mutual externality; outside the ego, in pure space, mutual externality without succession . . . since succession exists solely for a conscious spectator who keeps the past in mind . . . ' (TF, 108 – 9). (Cf. TF, 227.) By the time of *Matter and Memory* he was ready to attribute to things 'a real duration and a real extensity', seeing the source of difficulty in 'the homogeneous space and time which we stretch out beneath them' (MM, 281 – 2). He did not himself seem to see any inconsistency between the two books; in the summary of his thought at the start of chapter 3 of *Durée et Simultanéité* he refers in footnotes to both books indifferently as sources for his views – though admittedly he refers ambiguously to the 'development' of his views therein. However, there is clearly on the face of it a change of view, so let us ask why. The answer is not simple, and will lead us eventually to other parts of his system, notably to the relations of mind and body.

Let us return to the melody example. A melody is essentially a unity, which owes its whole nature to change. It has to last its whole length to be the melody it is, while a person can be cut off in his prime without ceasing to be, or to have been, that person, even if he did not reach his full potential. Times cannot coexist, so if something essentially involves different parts of itself existing at different times how can it ever exist together as a whole, and so how can it exist in the external world at all? The only way, it seems, in which a melody can exist is for it to be held together by something that can unite its different temporal parts, and the one thing that seems capable of doing that is consciousness. A melody is essentially something experienced. What exists in the world outside is simply the different notes, or perhaps the different vibrations that underlie the notes. If we take this idea to the limit we shall find that what exists outside in space is something momentary. This is the cinematographic view of reality for which, or rather for his opposition to which as the ultimate description of the world, Bergson has become so famous. The spatial world is a set of momentary cross-sections of reality, which in the limit are infinitely many and form a continuum. Space is infinitely divisible. But this means that the spatial world is in danger of falling apart. The world the mathematician deals with he calls 'a world that dies and is reborn at every instant, – the world which Descartes was thinking of when he spoke of continued creation' (CE, 23-4). The world Descartes was thinking of was undoubtedly the real world. (See e.g. his *Principles of Philosophy* I §21.) But we need not beg questions here about how far 'the world the mathematician deals with' can be identified with the real world, i.e. questions about whether for *Creative Evolution* the real world does not also contain duration.

It is here that we must bring in the fundamental distinction that Bergson draws, between movements and trajectories. (See especially MM, 246 – 50.) When I move my hand I move it through a space, which is infinitely divisible and contains infinitely many points. But the movement itself is a single movement (we are considering a simple continuous movement in one direction) and is not divisible. Bergson seems here to have two arguments for his conclusion. The first is that when we look at the movement of our hand we seem to be able to break it up according as we consider the various points in space at which it might, though it doesn't, stop; but kinaesthetically the movement appears to us as a single and indivisible phenomenon. Taken by itself this argument would be pretty weak, applying only to those movements we had kinaesthetic awareness of. But Bergson claims that by considering this case we can see that sight too really takes in the movement as a single whole. He presumably means us to infer that it does so in other cases too, so that the choice of example is merely suggestive. The reason we are misled is because we forget that in considering the possible stopping-places of the movement our mind momentarily halts at them, and therefore we attribute a similar momentary halting to the object, though in fact it does not halt at all, even momentarily: 'we must not confound the data of the senses, which perceive the movement, with the artifice of the mind, which recomposes it' (MM, 247). The argument still only applies to movements we are sensible of. Elsewhere he asks us to consider watching a shooting star, where we shall find 'a natural and instinctive separation between the space traversed, which appears to you under the form of a line of fire, and the absolutely indivisible sensation of motion or mobility, (TF, 111 – 12). But this is to bludgeon us with selected examples. So far as the argument is an empirical one, as it appears to be, will it apply also to watching the minute-hand of a clock, or the hour-hand? There is a difference here from the melody case, for a melody, even that of a funeral march, has a structure, and we can reasonably say it must reach its end if it is to be that melody at all, as we said before; but this does not seem to apply to any of the cases we are now discussing, at any rate without further argument. If my hand, or the shooting star, had stopped sooner might we not have had the same movement only shorter? It is indeed true that the movement must have some length in time, but it is not so obvious that it must have just the length it does have, or cover the distance it does cover, and so that it must have the sort of unity that a melody has.

The second argument asks, 'how should a *progress* coincide with a *thing*, a movement with an immobility?' (MM, 248). This has the merit of not being confined to movements that are perceived, but it is ambiguous as it stands (the French has 'comment . . . coinciderait'). If it means how could they coincide, we can simply ask, why shouldn't they? If it means why must they coincide, i.e. why should we expect them to, the point is presumably that they are widely different sorts of things. But as an argument this would beg the question, which concerns just *how* different they are. It could at most have a certain persuasive force. Bergson pushes the argument a bit further when he insists that duration cannot contain instants in the way that space contains points, just because of this unity of any change, though the apparent analogy with space tempts us to postulate such instants. 'While the line AB symbolizes the duration already lapsed, the movement from A to B already accomplished, it cannot, motionless, represent the movement in its accomplishment nor duration in its flow' (MM, 250). The motionless line cannot be what moves, but the movement itself does not move and it is less clear why the line cannot 'represent' it. The movement has a direction, which the space traversed hasn't. Would it help to barb the line and make it an arrow? Perhaps he would say that a line could be regarded as made up of little lines, while the arrow is not made up of little arrows: each little arrow would need its own barb. But this is a mere feature of the symbol chosen. One could have represented the direction by systematically varying the colour or thickness of the line and then the line could be regarded as made up of similarly variegated little lines. It is true that a point could no longer be variegated in this way, and so could no longer represent the direction; but for that matter it could not represent the extendedness of the original trajectory either. And again, if the movement had stopped sooner, why shouldn't we say we had the same movement, only shorter, just as we would have the same trajectory, only shorter? Of course if the movement is defined by its beginning and ending a shorter movement will not be the same movement; but the same applies to trajectories.

5. Zeno

Before going any further let us look at how Bergson uses his distinction to solve Zeno's paradoxes, one of the main stimuli to his whole philosophy (Q, 1542), and a topic he returns to again and again in his works, particularly in chapter 2 of *Time and Free Will*, chapter 4 of *Matter and Memory*, and chapter 4 of *Creative Evolution*.

Zeno is said to have produced forty paradoxes, of which a handful survive, usually not in Zeno's own words. Their interpretation is controversial but they seem in general to aim at showing that any kind of plurality or motion is incoherent, and so a mere appearance. We are concerned with four, the most famous ones, usually called the paradoxes of motion. They can be briefly described as follows. The Dichotomy: you can never finish a journey, because first you must cover the first half, then the third quarter, then the seventh eighth, and so on; there are infinitely many things you must do before finishing your journey. Indeed you can never even begin it, since before covering the first half you must cover the first quarter, and before that the first eighth, and so on; you must do infinitely many things before completing any bit of the journey you care to mention. The Achilles: Achilles races against a tortoise, whom he generously but fatally allows to have a start. First he must reach the tortoise's starting-point. But while he does so the tortoise has moved on to a new position, so next he must reach this new position. But while he does this the tortoise has again moved on, however slightly, to a third position, so Achilles must next reach this, and so on; he must do infinitely many things before overtaking the tortoise. The Arrow: at each moment of an arrow's flight it is at a certain place, exactly equal to its own length – at no moment can it occupy a place greater than its own length. Therefore there are no moments left at which it could move from one place to another, so it can never move at all. The Stadium (as Bergson calls it), which is hard to state without begging questions about its interpretation, so I give Bergson's own statement of it:

> Let there be a moving body which is displaced with a certain velocity, and which passes simultaneously before two bodies, one at rest and the other moving towards it with the same velocity as its own. During the same time that it passes a certain length of the first body, it naturally passes double that length of the other. Whence Zeno concludes that 'a duration is the double of itself'. (MM, 252n.)

Bergson applies the same solution to all of them, the distinction we have just been considering between a movement and its trajectory. The trajectory is indeed infinitely divisible but the movement is not. But the movement is the only thing that has to happen in time – the trajectory doesn't 'happen' at all – and so the movement simply occurs as a single unity. It may *take* time, but this does not interfere with its unity as a movement, for duration, as we saw, does not contain instants in the way that space contains points.

It is fairly clear how this solution applies to the first three paradoxes. The Dichotomy simply gives us a harmless division of space, and when we introduce movement over that space the Dichotomy has no application to it. Achilles moves until he catches his tortoise, no matter what we say about the ground over which he moves. The Arrow similarly flies through the air and never *is* at a point at an instant, because there are no instants. How the solution applies to the Stadium is not quite so clear. Although Bergson calls it 'the most instructive . . . perhaps' of the four, he actually says rather little about it and confines it to a footnote (MM, 252 – 3). I will defer it for the moment.

The solution raises various questions. First a point of interpretation. Why are we given as a reason why Achilles outstrips the tortoise that the steps of each of them 'are indivisible acts in so far as they are movements, and are different magnitudes in so far as they are space' (TF, 113)? Surely the point was supposed to be that the steps were movements and *not* themselves spatial magnitudes, though they covered space? But the point of distinguishing movements from spaces was that spaces could be infinitely divided – could be, not that they must be. The space covered by each step can indeed be divided *qua* space, but not *qua* magnitude, for when it is divided it ceases to be the magnitude of anything, or at least of anything relevant. The point of mentioning the magnitudes is simply that they can be added, and the sum of the magnitudes of Achilles' steps exceeds that of the tortoise's.

However, this raises a problem. If the magnitudes can be added to give a total which is itself a magnitude of something relevant, Achilles' total journey, why when divided do they not give a magnitude, not just of something irrelevant (someone else's journey that happens to be of that length), but of something relevant, a part of Achilles' journey? For surely if the steps he took are parts of his total journey the sub-steps corresponding to the divided magnitudes must be parts of the steps?

Bergson might offer two answers to this: first he might say that it is only infinite divisibility that produces the paradox; there is no reason why Achilles' steps should not be divisible into sub-steps provided these are only finite in number. But this reply, whether or not adequate in itself, would go against the spirit of his general position, which emphasises an asymmetry between journeys and distances. If we allow sub-steps to exist corresponding to the sub-distances we could as well point out that the distances too cannot actually be divided to infinity; whatever will be true of the sub-steps will be true of the sub-distances and vice versa.

The answer I suspect Bergson would really give is this: when we add the spaces covered by the different steps we get a space that is indeed the magnitude of something, namely of Achilles' complete journey. But when we divide the spaces in an arbitrary way the resulting spaces could be the magnitudes only of parts of steps – but steps don't have parts, so the spaces are not the magnitudes of anything (or anything relevant) after all.

This answer leads to a second question raised by the original solution: when do movements have parts and when don't they? The journey evidently has the steps as parts, for when the space of the journey is divided in a certain way the resulting sub-spaces are magnitudes of something, namely of the steps. Yet the steps do not themselves seem to have parts, and certainly not arbitrary ones. So why do the steps constitute parts of the journey?

Bergson makes clear that 'each of Achilles' steps is a simple indivisible act' and is 'of a definite kind', and contrasts the situation with one where two tortoises race each other but 'agree to make the same kind of steps or simultaneous acts, so as never to catch one another' (TF, 113). What exactly are the tortoises supposed to be agreeing to? Not to travel at the same speed, for Bergson presents this picture as being the way Zeno in effect regards the situation, and Zeno deliberately chooses runners who do not travel at the same speed. Call the two tortoises A and B, with B starting ahead. What they agree is evidently this: A will first journey to where B was when they both started, then to B's position at the end of the first stage, then to B's position at the end of the second stage, and so on. Meanwhile B will journey in corresponding stages, one to each of A's, i.e. each of B's stages will begin and end simultaneously with the corresponding one of A's. But in a sense this is something that will happen anyway, as Zeno would insist. Why do they have to agree on it? Bergson would say they don't, but that each stage of either of their journeys must be a separate action, and agreement is brought in just to make this clear. If one can agree to do something then the something must be an action. When Achilles runs, by moving his feet in the normal way, each step can be regarded as an action because it is something which at least in principle he could decide on. Each of his steps could be preceded by a decision, and in so far as he is monitoring his own progress, intending to stop when he has overtaken the tortoise, it might be said that in an etiolated sense it is so preceded. No doubt he *could* decide to take the first half of such a step, or the first quarter, but what he certainly couldn't do is take infinitely many such decisions, one for each of the infinitely many bits his trajectory could be divided into.

On this view movements are being individuated as those stretches of activity that are governed by a single mental act, of decision etc. To find out how many movements Achilles made perhaps we need do no more than ask him how he achieved his goal. He will probably say something like, 'First I moved my left foot forward, then my right foot forward, . . . '. He is unlikely to say 'First I moved my left foot forward one inch, then a second inch, . . . '. In a sense he did do that perhaps, but that is not how he looked at it; that would not describe his actions (plural).

But what about those movements that are not actions in this sense, not governed by a unitary act of mind? This is an important question since Bergson emphasises so much throughout the connexion between duration and consciousness. So far as Achilles' steps go he could say that if, as seems plausible, Achilles paid no attention to these but simply concentrated on the single task of catching the tortoise, putting his body, as it were, into the right gear and then letting his legs do whatever was necessary, then his total journey was a single movement and did not after all have parts.

But this does not take us very far. We have only to look at the Arrow paradox to find a movement which clearly has no connexion with mentality and is not an action. Neither Bergson nor anyone else is interested in the fact that arrows are normally fired by archers. The attempt to individuate movements by reference to mental acts governing them was suggested by the apparent relevance of what the tortoises agreed on, and also by the connexion of duration to consciousness. It may or may not be significant that *Time and Free Will* doesn't mention the Arrow, but we will return later to the relations between *Time and Free Will* and its successors. *Matter and Memory* too gives only a single sentence to the Arrow, where it is attributed to the fallacious attempt to distinguish indivisible moments in the duration of its movement (MM, 252). For *Creative Evolution*, however, the Arrow becomes the main paradox (CE, 325 – 7). The diagnosis is the same in effect as in *Matter and Memory*, but now we have an explicit criterion for individuating movements without reference to mentality: the only thing that would justify saying the arrow ever 'is' at a certain point is if it stopped there; but then we should have two flights separated by an interval of rest. 'A single movement is entirely, by the hypothesis, a movement between two stops; if there are intermediate stops, it is no longer a single movement' (CE, 326). There is a certain vagueness about this. If a movement abruptly changes direction, or velocity, does this count as a 'stop' of the original movement? We can divide Achilles' movement provided we respect its 'natural articulation' (CE, 328), i.e. presumably its steps; but Bergson can hardly suppose that Achilles stops dead after each step. And what if the change is continuous, as in curved motion, especially if it is also unsystematic, as in the flight of a bird? We could be tempted to think here of a sort of causal theory of individuation: a movement is a single movement if it has a single cause, and a second movement occurs if some further force has to intervene, as when a change of direction occurs. But this would certainly not do as it stands, since Bergson evidently thinks of a motion which starts and then stops (the flight of an arrow to, but not beyond, its target) as a single motion. The stopping, which requires a second force (the target's resistance), doesn't count as a second movement. (See also MM, 256 – 7.) He might reply that a single movement can indeed be divided, as Achilles' journey can be divided into its steps. What is important, and needed to answer Zeno, is that it cannot be divided indefinitely merely because its trajectory can. This would allow the arrow to have a single motion, but at the expense of taking away our reason for saying that it has. It would not cease to be single merely because it was divisible, but we could not offer its having a single cause as the ground for its being single.

Bergson goes on to discuss how Zeno's troubles might apply to what he calls 'qualitative becoming' and 'evolutionary becoming'

(CE, 329). It becomes evident here that what he is really attacking is a view that analyses movement into a set of rests. (Cf. Čapek 1971: appendix I, part II chapter 11, and Gale 1968: 388 – 92.) Zeno asks when the arrow manages to get from one position to another, since it uses up all the available instants by being *in* positions. Suppose one were to answer that movement just is being in different positions at different times, so that a mover is in different positions at each time within the relevant period, and the nearer two positions are to each other the nearer are the times at which the mover is in them; this is only a rough and ready definition, but should give the general idea. Such a mover does not remain in any position for any period, and there is no problem about how it gets from one position to the next since the positions form a dense set and there is no next; Bergson's cinematographic image is unfortunate here because cinematic reels do have adjacent frames. How the mover gets from one position to another is by occupying all the intermediate positions at relevant instants. The word 'is', and also the word 'occupy', for that matter, may cause some confusion here. 'Is' often means 'exists', and a thing can hardly exist simply at an instant, at any rate if it is a thing that exists in time as material objects do, and is not an abstraction like midnight. An ordinary object must exist for a period, not *just* at an instant. This suggests that if a thing's existence over a certain period is analysed into a set of existences at instants during that period then its existence seems to be analysed out altogether. Bergson may have in mind this idea that to exist at all, whether moving or resting, is to exist for a period. However, that something cannot exist *just* at an instant does not imply it cannot exist at an instant, though if it can existence loses the self-sufficiency Hume would desire for it.

Bergson sometimes expresses his view by pointing to the absurdity of saying 'that movement is made of immobilities' (CE, 325), and insisting that 'rests placed beside rests will never be equivalent to a movement' (CE, 329). If this is meant to rely on the fact that 'immobility' and 'rest' are opposites to 'movement' it smacks more of rhetoric than of sound argument. A brick is not a house but houses can be made of non-houses quite happily, and if contraries rather than contradictories are wanted a big heap can be made out of small heaps. And 'is' may denote correspondence, in the sense that to say that an object is at a place at an instant is to say that in a graph representing its history that place corresponds to that time. There need be no implication that it somehow endures at the instant, nor that it somehow manages to exist without enduring. What is true is that it could not be only at that instant that it was at some place or other.

However, Bergson may have a different and stronger point in mind. He talks of movement being 'made of' immobilities and of rests being 'placed beside' rests. Now 'made of' is ambiguous. It might mean 'consists of' in the sense of 'contains, and contains nothing else', 'is exhausted by'. Then a movement might well be made of immobilities in the following sense: the period of the movement contains instants and nothing but instants, and at any instant the object is not moving, in the sense that it does not at that instant get from one place to another – which is not to deny, incidentally, that it is moving at that instant in the rather more sophisticated sense in which we talk of something's velocity at an instant. But there is a more dynamic and indeed more literal sense of 'made of', whereby if movement were made of immobilities this would mean one could start with immobilities and so manipulate them as to end up with a movement, or that at least in principle one might have done so. But this one could not do, even in principle. One could not start somewhere and add immobilities, or place rests beside rests, and construct a movement. We might say then that a movement can be analysed into immobilities (in the first sense) but cannot be constructed from them, and movement might be called prior to immobility for this reason. That Bergson did not explicitly appreciate all this, though, is suggested when he refuses to allow that duration can in any sense 'contain' instants.

Actually modern physics could supply Bergson with a further argument, if his target is taken to be the view that motion can be analysed into instants at each of which the body is at rest. A moving body shrinks in the direction of motion, so a strictly momentary 'snapshot' of a moving body would show it as shorter than a similar 'snapshot' of a stationary body. In this sense therefore a moving body would be 'really' moving even at an instant, though only relative to frames other than its own.

One misleading argument can be noted in passing because of the conclusion it is used to support. Bergson finds difficulty in saying 'The child becomes a man' because the child cannot yet be called a man and whatever can be called a man is no longer a child. Bergson thinks we ought instead to say 'There is becoming from the child to the man', where 'becoming' is the subject, which 'comes to the front' and 'is the reality itself'; thus we seem to find a linguistic support for Bergson's ontology (CE, 329 – 30). But the support is illusory. The child, while a child, cannot yet be called a man – but he can one day. The person whom it will be correct to call a man is the same person as the present child. The only reason he will not then be the same child is that he will not be a child at all. But the child will not have perished. 'Child' is a phase-term, not a sortal; it is a term describing a phase of something's existence, not describing the sort of thing it is. Bergson, no doubt naturally enough for the time when he was writing, ignores linguistic points of this kind.

Let us return to the Stadium. Bergson's treatment of it brings in a feature we have not yet mentioned. The interpretation he adopts is one usually regarded as unflattering to Zeno, who has simply confused absolute and relative motion: of course a moving body will in a given time pass twice as much of a body moving in the opposite direction as it will of one at rest, but what is surprising about that? It only seems surprising, the interpretation runs, if we ignore the fact that the two bodies the original mover passes have themselves different motions. For Bergson Zeno is committed to his conclusion for the same reason that he fails to solve his other paradoxes: he ignores duration and considers only the trajectory. 'How then', Bergson asks, 'should the two lines traced by the same moving body not merit an equal consideration, *qua* measures of duration?' (MM, 252n.). Why two lines? Presumably they represent the trajectories the original mover would cover in the frames of reference of the other two bodies, each considering itself as stationary. The solution, for Bergson, is to take account of the one real absolute duration, in the frame of reference of which any body will have just one trajectory (of zero length if it is at rest). The new feature I mentioned a moment ago is this idea that duration is absolute.

6. *Absolute and relative. Absolute motion*

To say that space or time is absolute can mean one of two things. (Cf. H.M. Lacey 1970 for some discussion of the relations between the two.) It may be contrasted with the Leibnizian view that space and time are relations between the things commonly said to be 'in' them and all motion is relative. For the absolutist space and time provide a framework prior to their contents and in principle there can be space void of objects and time void of events. Or the contrast may be rather with Einstein's view that the spatiotemporal relation between things may differ according to the frame of reference one adopts, not just in the trivial sense that whether A is to the left of B depends on where one views them from, but in the sense that the length of a body or the temporal order of two events may differ for different observers. On the Leibnizian question one could in principle take different views for space and for time, and one cannot assume that Bergson must necessarily take the same view on each. Einstein's theory on the other hand must be accepted as a package if at all, at least in this respect, since it ties space and time together into space-time. Bergson could hardly be expected to express an explicit view on the Einsteinian question in his first two books, or even in *Creative Evolution* (published in 1907), though later he regarded Einstein's theory as compatible with, and indeed supporting, his own, at least to some extent; more on this later. But let us pursue the Leibnizian question for the moment.

In the footnote on the Stadium Bergson speaks of duration as 'a kind of absolute' of which a *determined* portion elapses (italics his). At the same time it is 'in consciousness or in something which partakes of consciousness' (MM, 252n.). But there are many consciousnesses, and the rhythm of duration may be different for each: 'there is no one rhythm of duration' (MM, 275), and 'that homogeneous and impersonal duration, the same for everything and for every one, which flows onward, indifferent and void, external to all that endures' is, as we saw, 'an idol of language' (MM, 274). It is this second-rate homogeneous duration, time rather than duration proper, which would provide an absolute framework if anything did. (But cf. CM, 212 – 13 (180).) So in what sense is duration absolute?

In his 'Introduction to metaphysics', published in 1903, Bergson distinguishes two kinds of knowledge, which he calls analysis and intuition. We shall discuss these later, but he introduces them by taking movement as an example and asking how one might speak of it as absolute. In fact, as he tells us later, he was looking for a definition of 'relative' and 'absolute' (M, 731). The conclusion he comes to is this: Movement as studied from the outside is relative, for our description of it will depend on whether we ourselves are moving and on what system of coordinates we adopt. To speak of absolute movement 'means that I attribute to the mobile an inner being and, as it were, states of soul; it also means that I am in harmony with these states and enter into them by an effort of imagination' (CM, 187 (159)).

It is perhaps not entirely surprising that this was misunderstood, at least in Bergson's opinion. In a letter to a journal in 1907 he complains that a certain Le Dantec in a review of *Creative Evolution* had misunderstood him in at least two ways, and sets out to clarify his position accordingly (M, 731ff.). He does not deny that when sitting in a train one can be deceived about whether it is the train or the platform that is moving. It may be with this in mind that he added a later footnote to the 'Introduction to metaphysics' saying that he was not offering a criterion for when a movement is absolute or not but simply defining *what one has in mind* (his italics) in calling it absolute 'in the metaphysical sense of the word' (CM, 305 note 21 (Q, 1393)). One might ask in this case how one can be *deceived* about whether the train is moving. He could reply that one normally thinks of some frame, e.g. that provided by the earth, as absolute, not in a metaphysical sense but in the sense that one treats it as basic for scientific or practical purposes, and that one can be deceived about whether the train is moving relative to that frame. The second point is more obscure. Le Dantec had taken him to mean that one could only get to grips with a movement by imagining that one is oneself the mover. Bergson asks who before Le Dantec had ever thought of such an extraordinary method, displaying an asperity that Le Dantec hardly deserved, one would think; how else might one 'enter into' the 'states of soul' of something, even 'as it were' states, 'by an effort of imagination'? The only gloss on his view that he gives here is that 'one may well suppose that the internal sensation of muscular movement lets us penetrate further into the inner nature of movement than visual perception of external displacement'. But he does quote a favourite passage (he quotes it at MM, 256, and in *Durée et Simultanéité* (M, 87)) from Henry More, who commented on Descartes's relativistic account of motion: 'When I am quietly seated, and another, going a thousand paces away, is flushed with fatigue, it is certainly he who moves and I who am at rest.' (Translation from MM, 256; see More 1679.)

One might think that this amounted to defining real or absolute motion in terms of the forces that cause it (muscular effort in More's example). But *Matter and Memory*, though insisting that real movement does indeed exist, repudiates this approach (MM, 256 – 7). Force is known only by the spatial movements it allegedly produces, and so we are led back to absolute space, which science usually tries to avoid. (Cf. here again H.M. Lacey 1970; I will return to absolute space below.) Alternatively, he goes on, we might bring in 'profound causes, analogous to those which our consciousness believes it discovers within the feeling of effort' (ibid.); but he thinks such feelings can be reduced to consciousness of movements, so that such a causal account of movement would be circular. Whether or not we accept this James/Lange-like account of sensations of effort we might well feel that such an account could at most tell us where psychogenetically we get the idea of absolute movement from and could not give us an analysis of absolute movement itself.

In his own argument for absolute motion Bergson first distinguishes geometry, for which all movement is relative, from physics, and then simply points out that if there were no real motion nothing in the universe would change (MM, 255). Descartes and others have treated every particular motion as relative while treating the totality of movements as an absolute. This seems to be a contradiction, but one might wonder whether it is. If a car passes me as I stand in the street and I choose to say that I passed the car there is a sense in which my motion is illusory and the car's motion is real. This is because we take the earth as providing a reference frame. We could of course say the opposite, though to do so might have disastrous effects on our mechanics and dynamics. Still, perhaps we are tough-minded enough to live with that. But what we cannot do is say that both my motion and the car's are illusory, at least on pain of joining the Eleatics in dismissing the world of appearance altogether. In this sense motion as a whole is real, and if we are to describe it we must adopt a frame wherein *some* particular motions count as real. But nothing in this seems to compel us to adopt one frame rather than another, however much extraneous reasons from mechanics or the pursuit of scientific viability may persuade us to.

I am not clear whether Bergson would agree that it is not contradictory to say that every particular motion is relative but that motion as a whole is absolute. In a way it is just what he wants to hold himself, and *Durée et Simultanéité* insists that variation in distance is real, though the movements from which it results are relative; but that movement is absolute is a further hypothesis (M, 87 – 9). I have suggested that the sense in which motion can be absolute even if every particular motion is relative is that some particular motions must be *treated* as non-relative, though it doesn't matter which (at any rate within certain limits; if we choose one motion we may be committed to choosing certain others with it; but let us ignore these complications). Bergson, however, makes the position depend on a distinction between geometry and physics. The reason Descartes treated particular motions as relative while formulating laws of motion as though motion were absolute was that he 'handles motion as a physicist after having defined it as a geometer' (MM, 255). Bergson then goes on to conclude that motion cannot really be spatial. A step as radical as this, a step which is important for his own philosophy of motion, suggests he was rather seriously impressed by the apparent contradiction we are discussing.

Bergson, then, holds that motion is not really a change of position but is rather a change of state or of quality (MM, 258). This has implications for his ontological views on substance and objects and on the relations between these and movement, which we shall meet later. But first some other questions about motion, time, and space, and before those an apparent inconsistency in Bergson's thought. Physics treats motion itself as against abstract descriptions of it (MM, 254 – 6). But elsewhere physics and geometry can no more get hold of the reality of duration than can any intellectual study. (See CE, 9 – 10, 355 – 9, M, 87 – 92.) It is true that *Creative Evolution* concerns duration while *Matter and Memory* concerns motion, but there does seem to be a real tension between them, since the *Creative Evolution* view is closely linked, as we shall see, with the *Matter and Memory* view that language 'freezes' the world (MM, 274). For *Durée et Simultanéité* physics too makes all motion relative, though it is not until general relativity theory that Descartes's programme for doing this is fully realised (M, 870 – 92). Only for metaphysics and psychology is motion absolute. Perhaps 'physics' in *Matter and Memory* means any study of the world but in *Durée et Simultanéité* only a mathematical rather than metaphysical study of it. But let us return to motion, time, and space.

7. Absolute space

Bergson holds that there is absolute motion. If such motion is spatial then there should be absolute space. But is there? In *Time and Free Will* his view is not altogether clear. He starts by saying, 'We shall not lay too much stress on the question of the absolute reality of space: perhaps we might as well ask whether space is or is not in space' (TF, 91). After this rather coy opening he goes on to elaborate the views we discussed earlier and does not explicitly address the question whether space is absolute. But the implied answer is Yes, in that he goes on to praise Kant for 'endowing space with an existence independent of its content' (TF, 92), adding that, 'Far from shaking our faith in the reality of space, Kant has shown what it actually means and has even justified it' (ibid.). We need not ask how far this avowedly controversial interpretation of Kant is valid as such. But what Bergson evidently means is that spatial relations cannot be reduced to or derived from other aspects of our experience; an act of mind is needed for our experiences to become spatialised, and that consists in 'the intuition or rather the conception, of an empty homogeneous medium' (TF, 94 – 5). This may not seem what is ordinarily meant by calling space absolute or real; it rather seems to make it mind-dependent. But it is an absolutist view in so far as it refuses to make space posterior to objects. The question of mind-dependence raises the general mind/body question, which we haven't yet come to, but we might say that for Bergson in *Time and Free Will* whatever reality the material world has is a spatial reality in the sense just sketched. In a passage I have already twice quoted Bergson contrasts what is 'within our ego' with what is 'outside the ego, in pure space' (TF, 108).

In *Matter and Memory* the picture is rather different. The notion of absolute space comes more explicitly to the fore, and Bergson rejects it. An absolute place, he thinks, would have to be individuated either by having some special quality or by being related to an absolute framework, which could only be given by the totality of space, which must therefore be finite. Neither of these ways will do. If space was heterogeneous, as on the first alternative, 'we should imagine an homogeneous space as its foundation' (MM, 256). Why should we? He does not say, but it looks as if he thinks that such qualitative differences between places could not be essential to them. Maybe one place was warm and another cold, but we could always imagine that the warm place became cold and the cold place warm, while *ex hypothesi* they remained the same places. This seems rather at odds with the passage I referred to earlier where animals were credited with an immediate qualitative appreciation of spatial directions or regions (TF, 96 – 7). But though that passage is puzzling I suggested earlier that it probably does not refer to directions in absolute space, and in any case need not refer to essential properties of those directions or regions. The second way of individuating absolute places was by letting space be finite. Bergson rejects this because a finite space would always be bounded by another space, and so would not be finite after all. This is false, as Bergson must surely have known; a non-Euclidean finite space need no more be bounded than the finite surface of the earth is. He should have said that given no other properties an unbounded finite space would have the same sort of symmetry as an infinite space and so would serve no better as a means for individuating places, while even a bounded one would only partly serve.

We should perhaps mention one further possibility, though Bergson could hardly have known of it until fairly late in life. Could we use the curvature of space to distinguish absolute places? Modern physics associates this curvature with the presence of matter. We need not bother whether the curvature of space at a given place just is the presence of matter there or whether it is something logically independent which causally follows from that presence. In either case it seems that we could only have absolute places by denying any movement, which none but the Eleatics would want to do. For if the curvature of space in some place just is the presence of matter there then if that matter moves the place will move, which hardly seems consistent with calling it absolute; and if the curvature results causally from the presence of matter then if the matter moves the place will alter its curvature and so cannot be defined by reference to it. A full treatment of this would take us into some further refinements, e.g. concerning rotation and movement in closed curves, but I think I have said enough to show that this approach is not very promising.

Bergson then seems to think as follows: space is not absolute. Therefore every particular motion is relative. But motion is absolute or real. Therefore motion is not spatial. I have argued that the relativist could hold the first three propositions without committing himself to the fourth by interpreting the third like this: We may be able to replace any particular statement of the form 'X is moving' by a statement of the form 'X is at rest', if we choose a suitable reference frame, but we cannot so treat all such particular statements at once. The relativist indeed could hardly deny the third proposition if interpreted like this. But Bergson interprets it in a stronger sense, and so draws his conclusion that motion is not, or not merely, spatial; the 'not merely' (MM, 256) presumably means that motion does indeed involve a trajectory, but also essentially has the qualitative nature mentioned above. We will return to this in chapter 4.

8. Absolute duration. The two stages in Bergson's thought

Now that we have discussed whether movement and space are absolute for Bergson let us return to the question in what sense duration is absolute. The question arises, as we saw, because 'there is no one rhythm of duration' (MM, 275); similarly he talks of comparing 'the various concrete durations' (CM, 212 (180)). While 'space is, by definition, outside us' (MM, 273) 'with duration it is quite otherwise' (ibid.). In 1891 he wrote of duration 'whose essence is to flow without ceasing, and so to exist only for a consciousness and a memory' (M, 353), and he ties the parts of duration strictly to the parts consciousness sees in it, or rather makes in it: 'if we distinguish in it so many instants, so many parts it indeed possesses' (MM, 273 – 4). We also saw earlier that *Matter and Memory* allows a real duration to things in themselves. Earlier in *Matter and Memory* he points to the enormous difference between the time-scale of microscopic physical events and the shortest intervals we are capable of detecting, and then, referring to this duration of physical things in themselves, he asks, 'what is this duration of which the capacity goes beyond all our imagination?' (MM, 272 – 3). He might well ask, since having seen duration tied so closely to consciousness we now have a kind of duration quite inaccessible to consciousness. The answer, so far as there is one, fills some fairly dark pages. He starts by distinguishing it firmly from both 'our' duration and the homogeneous impersonal time which is 'an idol of language' (MM, 274). It differs from 'ours' for obvious reasons, and from homogeneous time presumably because it is simply not homogeneous and cinematographic; being duration rather than time it must have the unity proper to duration. It is after dismissing this homogeneous time that he makes the remark about there being no one rhythm of duration, as though anyone who thought there was a single rhythm would be thinking of this homogeneous time, and much of the ensuing discussion is concerned to distinguish it from the various conscious rhythms of duration, i.e. to distinguish from each other the two things from both of which, he has just told us, the duration we are looking for must be distinguished. One clue comes where he says,

> if you abolish my consciousness, the material universe subsists exactly as it was; only, since you have removed that particular rhythm of duration which was the condition of my action upon things, these things draw back into themselves, mark as many moments in their own existence as science distinguishes in it; and sensible qualities, without vanishing, are spread and diluted in an incomparably more divided duration. (MM, 276)

Here in the absence of consciousness we have a reference to an 'incomparably more divided' duration, but not to an infinitely divided one. The criterion for distinguishing moments in it is appeal to science, and the content makes clear that what Bergson has in mind is the physical vibrations of light-waves etc. Here we have vast numbers of successive movements occupying a period which is undetectable by consciousness. If a one-second burst of red light were played, as it were, at a speed slow enough for us to be conscious of the separate vibrations it would occupy twenty-five thousand years (MM, 272 – 3). The duration is here being divided according to the movements that take place in it, and we are back to our old question of seeking criteria of individuation for movements. Vibrations provide a fairly good tool here, since they involve if not a stopping at least a clear reversal of direction of a movement. If we continued to divide duration (i.e. imagine it as divided) indefinitely, and in the abstract, without reference to any physical phenomena like vibrations, then we should have homogeneous time.

However, properly speaking duration cannot be divided. When you try to divide it thoroughly you end up with homogeneous time, as we have just seen. What duration can do is contain more or fewer events, and the only sense in which it can be measured is that we can count simultaneities in it, and Bergson defines simultaneity as 'the intersection of time and space' (TF, 110). (Cf. M, 354.) We measure movements by assuming certain events, e.g. ticks of a watch, as giving units of duration and then noting for each tick as it occurs the point along the mover's trajectory at which the mover arrives simultaneously with that tick. If the distances between successive points marked out in this way are equal then the velocity is constant. (Cf. TF, 117 – 19.)

Now our task of finding the sense in which duration is absolute, and thereby finding more about the nature of duration itself, is made harder by the discrepancy I referred to earlier between *Time and Free Will* and Bergson's later writings on whether duration occurs in the external world. Let us look at these two stages of Bergson's thought, starting with *Time and Free Will.*

9. The first stage

Science can only deal with time and motion by eliminating what is essential from them, duration and mobility (TF, 115), and 'Outside ourselves we should find only space, and consequently nothing but simultaneities, of which we could not even say that they are objectively successive, since succession can only be thought through *comparing* the present with the past' (TF, 116). He then offers a reason why science cannot take into account 'the interval of duration itself' (ibid.), and makes a rather remarkable claim: all the motions in the world might be speeded up, and if they were this fact would be quite unverifiable to science. At any rate 'there would be nothing to alter either in our formulae or in the figures which are to be found in them' and 'the same number of simultaneities would go on taking place in space' (ibid.). It is unclear what this claim involves. Would temperatures double (in degrees absolute)? And melting-points? And the speed of light? If mass remains the same but forces, which involve acceleration, do not, could this be unobservable? Nor will it help to adjust the mass. Altering the velocity will affect the momentum linearly but the kinetic energy exponentially, whatever we do with the mass. Could it be that Bergson is thinking of doubling the speed of a film and ignoring that the frames in the film do not causally affect each other in the way the events they represent would in real life? To take this further would involve Bergson's views on causation, which we will meet in the next chapter. Let us overlook these difficulties and allow Bergson his case. But we might then wonder what would be meant by saying that all things had speeded up, and how science could admit simultaneity but not succession. We will return to simultaneity later, but on the first question Bergson's answer is that consciousness would have 'an indefinable and as it were qualitative impression of the change' (ibid.). Later he repeats the point in a slightly expanded form (TF, 193 – 4). If a Cartesian demon speeded things up in this way there would be the same number of simultaneities for us to use in measuring, but the intervals between them would be shorter, and these intervals can never enter our calculations but are 'duration *lived*'; our consciousness would register the fact that they were shorter, but not necessarily register it immediately in these terms, i.e. that they were *shorter*, a quantitative term; it is not clear whether 'immediately' ('tout de suite', Q, 127) implies that consciousness would eventually see the change as quantitative, or if so how it would come to do so. But states of consciousness are processes, not things, and given some state we could not replace it by the same state only shorter. If we try reducing it we shall cut out some impression from it and so leave it qualitatively altered (TF, 196). To reduce an interval of time is simply to empty or impoverish the conscious states which fill it, and these conscious states provide the necessary basis for it to make sense to talk of shortening the time of all the other processes (TF, 197).

Bergson then is not just committing the fallacy of composition. He is not arguing that because we can conceive of any particular change as being speeded up therefore we can conceive of all changes as being speeded up. There is something common to the states of affairs before and after the speeding up, namely 'duration lived', and this is therefore an absolute.

To see more closely what this means let us take an example. Suppose the world goes on in its ordinary way for an hour and then the demon speeds things up so that in the next hour two hours-worth of events occur. This only makes sense if we can give sense to the phrase 'in the next hour'. This is where consciousness comes in. An observing consciousness is aware of the same amount of duration elapsing before and after the speed-up. It cannot use any external clock for this purpose. Nor could it use its perceptual experiences, since the things its perceptions would be of belong to the world, which is speeded up. It must just have a direct intuitive awareness of the passing of time, perhaps as we might think of ourselves as having during the silence of the night though without any heart-beats or metabolic processes to help us. Also we must presumably assume that the rhythm of our consciousness itself does not change.

However, we must not forget the constant association that Bergson makes between duration and heterogeneity, e.g. when 'duration within us' is described as 'a pure heterogeneity within which there are no distinct qualities' and 'the moments of inner duration are not external to one another' (TF, 226). The 'Introduction to metaphysics' tells us that,

> There is no state of mind, however, no matter how simple which does not change at every instant, since there is no consciousness without memory, no continuation of a state without the addition, to the present feeling, of the memory of past moments. That is what duration consists of. (CM, 211 (179))

(I have replaced 'mood' by 'state of mind' as translation of 'état d'âme'.)

This passage involves a major leitmotif of Bergson's philosophy, the survival of the past in the present and the role of memory. For the moment we are only concerned with its implications for duration. Two paragraphs after the passage I have just quoted we find that abstract time can only 'flow', i.e., presumably, become duration, by a continual change of quality. This leaves it unclear whether the change of quality is presupposed by duration or whether it is constituted by duration. Suppose I am feeling warm. Is the point that if there is to be duration the quality must change, e.g. I must feel progressively warmer or cooler? Or is it that the mere prolongation of this state of consciousness will itself qualitatively affect it, in the sense that to feel warm having just been feeling warm is intuitively a qualitatively different condition from feeling warm without just having been feeling warm? Much of what Bergson says about the essential temporality or durationality of consciousness suggests the latter. (Cf. CE, 2.) But there is a certain chicken-and-egg problem about this. Suppose I am lying awake at night feeling warm; I could be feeling other things as well, but for simplicity we will limit it to feeling warm. We are now supposing that if I go on feeling warm this fact alone will produce the qualitative multiplicity needed for my experience to involve duration. But would it be an experience at all if it did not involve duration? I think it is clear that a consciousness totally devoid of duration would not be a consciousness at all for Bergson. But then, if duration involves a prolongation of the feeling of warmth, what is it that it involves a prolongation of?

At this point Bergson might stage a counterattack. What would I or anyone else want to say about the relation between experience and time? Could anyone say that one could literally feel warm for one instant only? Surely feeling warm essentially involves time, just as essentially as feeling deep grief, to borrow Wittgenstein's example, though no doubt not as much time (Wittgenstein 1953: II i). To discuss this let us bring in a distinction made by Aristotle between what he called activities (*energeiai*) and processes (*kineseis*, which could also be translated 'changes' or 'movements'). (See *Metaphysics* IX 9, *Nicomachean Ethics* I 4.) Seeing is an activity, and 'I see X' is compatible with 'I have seen X' . Building a house is a process, and 'I am building a house' excludes 'I have built a house', unless another house is in question. It follows that there can be no first moment at which 'I see X' is true, or, as we might put it, the dense infinite set of moments covered by the activity of seeing X is open at its earlier end (at least). For Aristotle activities are 'perfect' or 'complete' at any moment they are going on, while processes are so only when they are over. My building a house is only complete when I have finished building it, but it is not true that my seeing X, or feeling warm for that matter, is only complete when I have finished seeing it or finished feeling warm.

Seeing X and building a house then both involve time, though in different ways. Can they be called prolongations of something? Building a house cannot, I think, but seeing X can, at least in this sense: for any period of seeing X that you care to take there is always a shorter period of seeing X partly coinciding with it but ending earlier. For anything you could call 'seeing X' there is something ending earlier which you could also call 'seeing X', though it is not true that there is something you could call 'seeing X' which ends earlier than anything else you could call 'seeing X'. So any feeling of warmth could be called a prolongation of something, without there being something of which it is a prolongation but which is not itself a prolongation of something. So Bergson could say that the question that ended the paragraph before last is an unfair one. In a sense duration involves prolongation of itself, for any duration could be seen as a prolongation of a duration – or rather of a possible duration – ending earlier.

To return to heterogeneity then, Bergson's view seems to amount to this: a continuing feeling of warmth – i.e. any feeling of warmth – will contain early parts which are preserved in memory in the later parts, and the longer it goes on the greater the amount of such earlier parts that are there to be preserved in this way, and so in this respect the experience is, and necessarily is, constantly becoming qualitatively different. Strictly one should not talk of 'parts' here, for they are not isolated or separately individuatable.

The whole point is that they merge into a continuous whole, and duration has no parts. But as we saw earlier language for Bergson necessarily distorts reality in the interests of intellect. The issue then, and the point where Bergson goes well beyond Aristotle, concerns the role of memory and the preservation of the past in the present. Bergson emphasises the non-momentary nature of experience. Ever since Augustine philosophers have been haunted by the thought that if we pursue the present it disappears without trace between the remorselessly encroaching past and future. (See Augustine 1964, and also Findlay 1941.) Bergson takes this thought by the horns and halts it in its track. It is simply not the case for him that we live one moment at a time.

Actually this could be developed in two ways. We might say that all our past experience was somehow still present with us, or we could say more modestly that experience is not momentary but always involves an ongoing stretch of time, a specious present, as it is sometimes called; we might compare cycling with a flat tyre, where a stretch of tyre, not just a point, is in contact with the road at any one time but no point on the tyre is permanently in contact with the road. There are various objections to the idea of a specious present, but for the moment let us just ask how it would serve the present purpose. On the former view (permanent survival of the past) we should be saying that feeling warm after having felt warm for a long period is qualitatively different from feeling warm after having felt warm for a short period; a continuing feeling of warmth is therefore continually changing in quality. On the latter view (specious present) we should be saying that feeling warm after having felt warm for a short period itself preceded by a short period of feeling warm is qualitatively different from feeling warm after a short period of feeling warm itself not so preceded; a feeling of warmth leaves a trace behind itself after it has disappeared. We will return to perception, memory, and the past in chapter 5.

So is duration absolute? Yes for Bergson in the sense that it is real. It is not a relation posterior to the terms it relates, and still less is it somehow illusory or put there by us. But it does not follow that there can be empty duration, duration with nothing happening in it, as for Newton there could be time with nothing happening in it. For duration, as it were, provides its own filling; for it to exist at all is for becoming to be taking place, but we do not have to have something independently definable which becomes.

10. The second stage

But if duration is real, and in this sense absolute, we come back to the question how it is related to consciousness and to the outer world. We have reached at least a partial conclusion on whether duration is absolute, but as I said before this question is not independent of Bergson's apparent change of view.

We have already asked what would be meant by saying that events in the outer world had been speeded up, and have concluded that the criterion is provided by consciousness. The problem was that it seemed at first that events in the world provided the only way of measuring the speed of other events or processes, so that to talk of doubling the speed of all events seemed to remove the basis for judging that such a doubling had occurred. Consciousness was brought in to solve this, but then it seems that consciousness itself does not provide a unique criterion because 'there is no one rhythm of duration' (MM, 275), a fact which is illustrated by a dream which seems to continue for days while the sleeper only sleeps for a few minutes. Presumably the point there is that the sleeper would experience those few minutes as much shorter than days if he were awake instead of dreaming. We fail to realise this (that there is no one rhythm of duration) because we have substituted an independent and homogeneous time for 'the true duration' lived by consciousness. The English words 'the true duration' suggest a contrast with other durations which are less 'true', and so suggest there is one true rhythm after all; but the French has 'la durée vraie' (Q, 342), where the definite article is less significant than in English, and the words may simply mean duration properly speaking, as opposed to homogeneous time. The point can only be that homogeneous time is more likely to be unique, or is more easily thought of as unique, than is duration.

We have already exposed, without reaching any very clear answer, the difficulty raised when Bergson tries to distinguish the duration of things in themselves from both 'our' duration and homogeneous time (MM, 274). Let us try another tack. Perhaps his point is that duration itself is simply a sort of 'togetherness' of things, and is incomplete until it is *given* a rhythm by consciousness. Until this happens it is incomplete in the way that a genus is until it is specified by differentiae. But just as there need be no one species that exhausts the genus, so there need be no one rhythm given by consciousness. Homogeneous time on the other hand is simply an infinite set of moments laid out together, whether the infinity is regarded as actual or only potential, and no question of its being gathered up into a 'rhythm' arises. The course of events could be said to have doubled its speed if every consciousness which was conscious of those events at all was conscious of them as going twice as fast. If only some consciousnesses were conscious of them as having speeded up we would attribute this to something about those consciousnesses themselves; they had just emerged from a state of ennui perhaps. Strictly this will not do as it stands. We would not want to refrain from saying that everything had doubled its speed simply because, although most consciousnesses were aware of things as going faster, a few happened at that moment to suffer from an onset of ennui which cancelled out this effect. But we need not pursue this sort of complication in the present context.

It is perhaps natural that if Bergson thought along these lines he started, in *Time and Free Will*, by denying that duration existed in the outer world. For how could it, if duration essentially involves consciousness and there is no consciousness in the outer world? Still, the situation is unsatisfactory, as Bergson himself saw later – at least in effect: as I said before, he did not seem to be conscious of having changed his view, though he plainly did. For our own part we can simply say how implausible it would be to say that without human beings or similar observers there would be no passing of time at all, and indeed properly speaking no events. One might wonder how the world got on 'before' life evolved. For *Time and Free Will* external things change, but their moments do not *succeed* one another, except for an observing consciousness. Like Zeno's arrow, one feels inclined to say, they may have changed but they never do change. They do not endure, but 'there is in them some inexpressible reason in virtue of which we cannot examine them at successive moments of our own duration without observing that they have changed' (TF, 227). Similarly there must be some incomprehensible reason in things ('en elles') 'why phenomena are seen to *succeed* one another instead of being set out all at once' (TF, 209 – 10). (As Čapek points out (1971: 194), 'incompréhensible' is left untranslated. So is 'en elles'.) This is not inconsistent with the other passage provided 'are seen to succeed' means the same as 'succeed one another for an observing consciousness'.

In *Time and Free Will* the reason is left uncomprehended, but later, as we saw, the world has 'a real duration and a real extensity' (MM, 281 – 2). *Creative Evolution* develops Bergson's position using the image of an unfolding fan whose flanges represent momentary states of the world, which in the limiting case could be all spread out together. (See CE, 9 – 10, 355 – 9.) Apart from the old objection that without succession things would have inconsistent predicates, it is unclear what the fan spreads out in, since the three dimensions of space are already used up in describing any one momentary state of things. In fact *Durée et Simultanéité* makes this point, arguing that a fourth quasi-spatial dimension would be needed, and that to spatialise time in this way ignores the reality of succession and would let us take arbitrary cross-sections of this four-dimensional manifold in incoherent ways (M, 195 – 204); he adds that this latter is just what relativity theory does, albeit not arbitrarily (M, 203). To return to the fan, there would also be a certain difficulty in unfolding it completely since its flanges, representing the different momentary states of the world, would have to be non-denumerably infinite. Perhaps we should regard it as a fan that could never be completely opened but could always be opened further than it had been so far. This might also free Bergson from the apparent fallacy of arguing that because the course of things could in principle be speeded up indefinitely therefore it could be speeded up infinitely. It is not so clear that the language of common sense similarly ignores the temporal dimension of things, as he claims (CE, 10). But we have already seen how Bergson thinks language has to 'freeze' the things it talks about. The application of the point is presumably this: since we have, according to him, no way of expressing duration, when we use a grammatically proper name like 'Smith' we are not really using it to name a continuing individual but rather as something like a variable ranging over Quinean time-slices of such a continuant – though if this *is* the point there is a difficulty about reconciling it with what I have just said about the unfolding fan: if the fan cannot be unfolded completely, can we any more easily have access to, and so name, completely momentary time-slices of our continuant? Yet, if we cannot, then whatever objections apply to our naming the original continuant should also apply to naming finite time-slices of it, since these will also be continuants. Does Bergson mean that we have access to some momentary time-slices but not to an infinite totality of them? More likely perhaps he does not think we have 'access' to any momentary time-slices (cf. MM, 75) but in naming something we have to freeze it and treat is *as though* it remained unchanging in the way that its name must be assumed to do if we are to use it for communication. Since only momentary time-slices would in fact involve no change we might express the point like this: when we use a name we pretend it names a momentary time-slice taken from a homogeneous set of such slices, any one of which could be named by the name in question.

However, Bergson continues, succession exists, and things do not all happen at once. Moreover they pass at some definite speed in the sense that I have to wait for something to happen, e.g. for a lump of sugar to dissolve when I put it into water. A similar point is in fact already made earlier. (See TF, 198.) The amount of waiting I have to do – which Bergson seems to think can be measured in terms of 'a certain degree of impatience' on my part – is not under my control but is forced on me from outside. Bergson concludes that the external succession, as against mere juxtaposition, has a real efficacity and so is objectively real, though this duration need not be identical with matter itself, or as he obscurely puts it, it need not be 'the fact of matter itself' (CE, 359). Bergson links all this with the notions of determinism and spontaneity (CE, 358), but we haven't reached that area yet.

11. Conclusions on the two stages

So now we have an external duration, which in the absence of Cartesian demons presumably proceeds at a constant rate, or, as we might prefer to put it, the processes in it each go on at a constant rate as measured against the background provided by the bulk of the others, except for where there is an ordinary causal explanation for them to do otherwise; strictly we could allow for a minority of them to fluctuate without a cause, but this is a subtlety we can ignore. But there are still problems. If, as I have suggested, objective duration is only generic, how can it constrain the melting of Bergson's sugar? And how can he experience this melting differently according to his mood or the rhythm he is living at? The answer to the second question is presumably that we must attribute such variations to changes in our own rhythm only if we can give some causal account of such changes, at least in principle. Given that we can identify changes of rhythm in this way, objective duration constrains the melting in that it is constant for any given rhythm; it need not be constant over all rhythms. But how is this objective duration related to consciousness? For Bergson insisted throughout his life that duration implies consciousness. (See the explicit statements in the late *Durée et Simultanéité* (M, 102) and in the admittedly secondhand report of a lecture he gave in 1901 (M, 515), as well as a review in 1891 (M, 353).) How then can there be duration in the outer world? This returns us to the whole question how far the later books do really diverge from *Time and Free Will*, where there is no duration in the outer world, and the answer lies in the different approach Bergson adopted not to duration but to the outer world itself. The reason duration implies consciousness is that it implies the survival of the past in the present, and this involves memory. The present chapter is not the place for a full discussion of the external world, but we can note the main point that it was only in *Matter and Memory* that Bergson really got down to studying the relations between mind and matter, and his approach there was to transcend the traditional dualism. Without begging questions about his attitude to realism and idealism we might say that he made matter rather more mind-like than it is usually taken to be, so that the notion of memory had some application to the material world. The fact that it was only in the later books that he came to develop a view on these questions may explain why he was less conscious than we might expect of any change in view. To say that the outer world does not contain duration and that it does plainly represents a clash of some sort, but the clash may not be so apparent as such if the change of view really concerns the nature of the outer world; i.e. Bergson could well go on holding that the sort of thing he had originally taken the outer world to be before properly examining it would indeed not contain duration. If duration is generic in the way suggested, it must receive its specifications both in ordinary human and animal consciousness and in whatever consciousness the world itself provides. But how these consciousnesses are related together is another question.

12. Criticism. The symmetry of space and time

Bergson's denial of the symmetry of space and time has been sharply attacked, notably by Berthelot (1913: 186 – 7, 350 – 6) and M. Boudot (1980). Well over half of Berthelot's book is devoted to a sustained attack on most of the leading elements of Bergson's philosophy, an attack based on a comprehensive reading of both Bergson and others but somewhat unsympathetic in nature; Bergson is not given much of a run for his money. The last ten of the book's 350-odd pages are devoted to extracting what is of permanent value in Bergson's philosophy, though even here much of the space is devoted to contrasting this with his errors. (See also 158 – 62.) His main virtue turns out to be that of isolating a psychological conception of time, his 'duration', in contrast to mathematical time. The main features of this duration are two. First, it essentially involves the notion of the present, and with it the past and the future, while mathematical time involves simply the notions of before and after. This reminds us of McTaggart's (1908) distinction between the A-series and B-series, which Berthelot, publishing in 1913, seems not to know of. Bergson himself had already published his main works by 1908, but never, I think, mentions McTaggart in his later writings. The spirit of his philosophy is that of a thoroughgoing A-theorist, agreeing with

McTaggart that the B-series by itself is senseless, but not that the A-series is incoherent. The second feature Berthelot attributes to duration is that, unlike mathematical time, it is not indefinitely divisible, and yet it does not consist of discrete elements. Berthelot then goes on to say that where Bergson goes wrong is in dismissing mathematical time as an illusion and contrasting duration instead with space. But in fact, Berthelot thinks, the same distinction between the psychological and the mathematical can be seen in the case of space – he refers to Berkeley and Malebranche for the psychological conception – so that what Bergson has done is treat as a contrast between time and space what is in fact a contrast that appears within each of time and space. For Berthelot then time and space are on a level, about as radical a challenge to Bergson as could be imagined.

Boudot is even fiercer, attacking Bergson's views on both the ontological status and the structure of space. Doubtless we can only apprehend a melody as a whole, but the same applies to a circle, he claims, and, as for the measurement of time being confined to counting simultaneities, in space too we can only count coincidences.

How far Berthelot is right in saying that Bergson takes a distinction that properly applies within each of space and time and uses it to contrast space and time themselves is not easy to answer simply. As Berthelot himself says, Bergson makes the distinction in the case of time – in fact it is his main claim to originality that he does make it – though he does dismiss one of the resulting terms (homogeneous time) as less real and ultimate than the other. In the case of space he makes a distinction that is partly analogous but only partly so, a distinction he could use to deal with Boudot's circle example. Boudot's point that coincidences play the same role in measuring space as simultaneities do in measuring time is a stronger one, since Bergson does see a radical difference between spatial and temporal measurement. He might reply that nothing spatial corresponds to the different rhythms of duration, though this itself is a none too clear notion. If we objected that spaces subjectively vary in size according to our distance from them he might reply that this was a different matter just because the varying depended on the distance in this way. But I suspect his main point would be that times, unlike places, are not there together. Strictly this is trivial because 'together' means 'together in time', but I will simply add that the following are among points that might tempt one to have some sympathy with him in not putting space and time on a level.

Time has an in-built direction, which space hasn't. We feel

tempted, however mistakenly, to say that time flows, but we feel no temptation to say space flows. Is this just because 'flows' is a temporal word? The M1 stretches from London to Leeds, but equally from Leeds to London, or between the two cities, and, while we often say things like 'Time flows', we don't often say 'Space stretches': why should we? To say that time flows is to say things change and pass away. We don't feel the same need to say things are distant, partly perhaps because we don't feel that distant places are permanently inaccessible in the way that past times are. We can move around in space in a way we cannot move around in time (even if whistle-blasts can: see Taylor 1955). If things are absent in space we can in principle do something to make them present. If they are absent in time we can do nothing. If they are future we must wait; if they are past we can but reminisce. Similarly two-way communication between people puts constraints on their temporal relations (their lives must overlap), but not on their spatial ones, assuming their temporal ones are suitably adjusted. It is not surprising then that we view the future in a radically different way from the past, and much of Bergson's philosophy is based on just this.

Two things can't be in (occupy) the same place at the same time; but neither can they be at (spend) the same time in the same place. Why then do we say 'You can't have two things in the same place' more naturally than 'You can't have two things at the same time', unless we specify: 'You can't have two things *there* at the same time'? Things are bulky and take up space in a way they don't take up time, apart from special contexts like 'time-sharing'. They endure through time, and would not be 'things' in the relevant sense unless they did. But, as we saw earlier, a thing is complete at any one time and has no temporal parts (though its lifetime has), whereas it is not similarly complete at any one place. Furthermore not all the things that exist in time exist in space. Beethoven's admiration for Napoleon had definite temporal limits, but it did not exist in space, except in the etiolated sense that it was Beethoven's and he existed in space. But it seems that all the things that exist in space exist in time. One could argue (Taylor presumably would argue) that most if not all of these facts are merely linguistic. But, even if so, they must have some basis, and there must be some reason why our concepts are as they are.

The symmetry position would again leave unexplained why most European languages at any rate, even if not languages like Hopi, have tenses, but do not have analogous devices for representing spatial relations. (See Whorf 1968.) The nearest device would be personal pronouns or the analogous inflexions of verbs, but the emphasis is only very partially, if at all, spatial, and the I/not-I framework splits apart from the here/there framework in utterances like 'I was there'. And, just as we have no spatial analogues for tenses, so we have none for phrases like 'no longer' and 'not yet'. It is beyond my scope here to discuss whether there is any connexion between facts like these and the fact that space has three dimensions and time only one.

A final point brings us back to Augustine's fear that the present was in danger of being squeezed out between the past and the future. No-one points to a similar danger in the case of the here. If I am now living in the present and not in the past or the future, I am also living here and not somewhere else. But where is here? Bromsgrove perhaps, where I happen to be writing? Or the room I am in? Or my body? But I am tempted to go no further. As an answer to the question where I am 'here' never means anything less than my body, or perhaps my brain, and there is no temptation to say that this is only a loose answer, as when I say that 'now' is Tuesday. In a way this has an ironic significance for Bergson, who wants to contrast homogeneous indefinitely divisible space with heterogeneous time that is 'held together' into duration. Now it seems to be time that is falling apart in the Augustinian way, while space is 'held together' into a sort of analogue of duration, at least in the case of persons. But perhaps it is the very process-like nature of time that makes us feel that only one bit of it can exist at once, while it is the homogeneous non-process-like nature of space that lets us relax our demands and allow space an unproblematic unity.

13. Bergson and Einstein

Some time ago I deferred the question of Bergson's relations and attitude to Einstein's relativity theory. Bergson's interest in Einstein was aroused by a conference paper by P. Langevin in 1911 (M, 131n.), and eventually he published *Durée et Simultanéité* in 1922. It has not usually been regarded as one of his more successful books, and did not originally appear in the centenary edition of his works because it was thought he had withdrawn it. However, it was eventually decided that, though he declined to publish further editions of it after 1931 on the grounds that his mathematical knowledge was too poor to let him defend it adequately, this did not amount to withdrawing it, and so it appears in *Mélanges*, as well as separately. (See M, x – xii, and also the editor's note at M, 1626 – 7.) One might add in defence of his mathematical prowess that it was sufficient in his youth in 1877 to win him first prize for an original solution to a problem in elementary mathematics proposed at a congress (M, 247 – 54). Bergson's main philosophical work in the relevant areas was completed well before 1911, so that *Durée et Simultanéité* with one or two short defences of it forms rather an afterthought, and needs only a brief discussion. (See M, 1340 – 7, 1432 – 51, CM, 301 note 5 (Q, 1280). For fuller treatment see Čapek 1971, especially part III chapter 8, and 1980, and Barreau 1973.)

Bergson takes up again Henry More's criticism of Descartes, which I quoted earlier, for making all motion relative. (See M, 87 – 92.) Within the confines of science itself, thinks Bergson, Descartes was quite right; only for metaphysics does motion become absolute, as we saw earlier in the discussion that so troubled Le Dantec. But Descartes's programme of making all motion relative, treating physics in a purely mathematical way as opposed to a metaphysical way, as we might put it, was not completed until Einstein's general theory of relativity. This was partly because ordinary language talks in absolutist terms and partly because of the practical usefulness of introducing a notion of force, as Newton did. But this led to an unhappy compromise where rectilinear motions of constant velocity were relative while rotation and acceleration in general were absolute. Only with the general theory of relativity did these latter become relative too.

So far then relativity theory, in its general form, brings to fruition what Bergson considered the proper line for physics, as against metaphysics, to take. In a long footnote, added to 'Introduction to metaphysics' in 1934, Bergson treated relativity theory as an attempt to give a mathematical representation of things independently of frames of reference, and so to give a set of absolute relations, which made the term 'relativity theory' rather misleading. His point is presumably to contrast it with the Newtonian system, which presupposes a particular frame of reference, given by absolute space and time, and he thinks philosophers have been misled by the name of the theory into thinking it makes science relative in the sense of deforming or constructing its object. (See CM, 301 note 5 (Q, 1280) and main text at CM, 45 (40).) In the footnote he says it cannot be used either for or against his own metaphysics of duration, but in *Durée et Simultanéité* itself he seem rather to think that it confirms it if anything, at any rate in the sense that the invariability of the speed of light, which the theory predicts, is what we ought to expect. The argument is rather obscure. (See M, 92 – 4.) Objects are subjective in the sense that our practical interests determine how we divide the world up. We find it convenient to treat a cat as one object and the mouse it chases as another (my example), rather than combining cat and mouse into a single object, say a 'couse'. Bergson infers that the motion of the cat, considered as affecting the cat in isolation (i.e., presumably, ignoring whether the earth etc. is moving), is equally conventional. Now we can no doubt say either that the cat approaches the mouse or that the mouse (unwillingly) approaches the cat, but what has this to do with the fact that we could talk in terms of a 'couse' instead? At any rate, Bergson continues, all that we can be sure of in such cases is that there is 'some modification in the universe', whose nature, and even its precise place, escapes us. Perhaps this means that we cannot say objectively even that there is a shortening of the cat/mouse gap, since we cannot say objectively that there is a cat and a mouse. Where we can talk of a movement which we know to exist in itself is when we feel a sense of 'effort' – excluding somehow, one presumes, cases like that of William James's paralytic who thought he had moved his arm when he hadn't. (See James 1890: vol.2, p.105.) But we can also talk of objective phenomena in the case of qualities, like colour, and their underlying movements (atomic vibrations etc.), and the phenomena of light fall under this heading. In the case of light we can talk of 'propagation', as opposed to the 'transport' involved in the relative movements of objects referred to above, and Bergson insists on the importance of the distinction. Since this propagation of light does not involve the transmission of particles he finally concludes that we ought not to expect it to be affected by or relative to human ways of perceiving and conceiving things, and so that we ought to expect its velocity to be constant in all frames.

The argument is complex and seems to amount to this: Common sense would say the observed speed of light should vary according as the distance between us and the light source is varying or not. Its real speed should depend on whether the light source itself is moving or not. If it is conventional whether the light source is moving the light can have no real or absolute speed apart from its observed speed. But for Bergson it is conventional whether the light source is moving but the light does, for metaphysical reasons, have a real or absolute speed. Therefore this must coincide with its observed speed. Now the argument bifurcates. If it is conventional whether the distance between us and the light source is varying then the observed speed, which is not simply conventional, cannot depend on this for any variations. But there is nothing else it can depend on. So it does not vary. If on the other hand it is not merely conventional whether the distance between us and the light source is varying, then if it is varying the real speed must vary according to whether the light source is itself moving. But whether the light source is moving is conventional, and so the real speed cannot vary in accordance with this. But there is nothing else for it to vary in accordance with. So it does not vary. But the real speed coincides with the observed speed. Therefore again the observed speed does not vary. But this is precisely what science, to the surprise of common sense, tells us does happen, and relativity theory involves it.

The bifurcation in the argument is a feature of my interpretation rather than of the argument itself. The argument thus interpreted seems valid. What it shows is that relativity theory involves asserting something strange but true (the invariance of the speed of light) that Bergson's theory also asserts. One might say that relativity theory harmonises with Bergson's theory. Relativity theory need not be the only scientific theory that involves the speed of light being constant, but that does not matter if it is the accepted theory. Whether the argument is sound depends on whether we can accept all its premises, and this is hard to assess because of difficulties in interpreting some of them, especially when we ask about the sense in which and the reasons for which light for him has a real or absolute speed. I will not discuss this further, but Bergson would be in a stronger position if he had predicted the invariance of the speed of light, i.e. drawn this implication from his own theory, before the appearance of the empirical evidence which seems to demand it. The Michelson – Morley experiment, which he makes much of in *Durée et Simultanéité*, was first performed (by Michelson) in 1881, as he tells us himself (M, 64). When he first learnt of it is not clear. Perhaps not until 1911. He does not, I think, mention it before *Durée et Simultanéité* in 1922. But there is no evidence either, I think, that he drew the implication in question before then.

The main reason why *Durée et Simultanéité* led to dissatisfaction in many of its readers, and perhaps ultimately in Bergson himself, lies in its treatment of the rocket paradox, originally raised by P. Langevin in his 1911 paper. The importance of this treatment goes beyond the paradox itself and concerns his whole conception of relativity theory. A physicist Pierre installs his twin Paul in a rocket and fires it off at a speed just less than that of light. At a certain point in space the rocket reverses its direction and returns at the same speed. Paul gets out, having spent two years (let us say) in his rocket – only to discover that his twin Pierre has spent two centuries on earth waiting for him. We take this fanciful example just for convenience, but the implication is that if we are twins and I leave you sitting in your chair while I go for a walk I shall return slightly younger than you. Exercise keeps you young, as we might say! Call this the weak paradox. The strong paradox arises when we note that nothing stops us yet from saying Pierre hurtled off with the earth, taking two years, while Paul waits patiently in his rocket for two centuries; in the first story Pierre becomes older than Paul, in the second vice versa, and both stories seem equally true.

The strong paradox is logically incoherent and no-one accepts it. To the weak paradox there are three relevant reactions. Some have argued that the two stories are indeed symmetrical, as the strong paradox maintains, but that the outcomes are the same, since whichever twin shoots off loses on the journey back what he gained on the journey out. While they are apart Pierre and Paul may differ in age, but this difference could only manifest itself via the transmission of signals, during which it would disappear. Therefore neither paradox holds. This account, however, no longer finds favour, and there is experimental evidence for the weak paradox. (See Rohrlich 1987: 67.) The standard response of physicists now, I believe, is to accept the weak paradox, arguing that only one of the two stories holds because only Paul suffers acceleration – Pierre would only do so if gigantic rockets were fitted to the earth. The relevant acceleration is that at the turning-point, not the blast-off and touch-down, which could be dispensed with by letting Paul be already travelling and flash past Pierre on both journeys. However, it seems to be agreed that acceleration itself has no direct effects. It simply ensures that Paul changes frames, which is what is needed. Space forbids a full discussion, but if acceleration is of no intrinsic importance it does not seem clear why Paul's rather than Pierre's frame-change counts as real, so providing the asymmetry needed for the weak paradox. Until some account is given of this the strong paradox still looms.

The third reaction is Bergson's own. Like the first it rejects the weak paradox, but for different reasons. (Bergson could not know the experimental evidence.) Einstein's account of the world is independent of frames of reference and so absolute, but abstract. Einstein's world is as real and absolute as Newton's, but it is a set of relations, not of things. (See CM, 303 note 5 (Q, 1283 = M, 1441).) To enter into the world we must adopt a reference frame, which may be Pierre's or Paul's but not both at once. Suppose we adopt Pierre's. Then, with Pierre, we experience Paul's absence as lasting two centuries, while for Paul, *as regarded by us*, it lasts only two years. But Paul as regarded by us is moving. As regarded by himself, however, i.e. within his own frame, he is at rest, while we are moving. Therefore he experiences the absence as lasting two centuries, while for us, *as regarded by him*, it lasts only two years. Paul as regarded by Pierre and Pierre as regarded by Paul are mere fictions. The real Pierre and the real Paul are each at rest in his own frame and experience the separation accordingly, as lasting two centuries. Moreover if we try to take a God's-eye view by adopting a third frame, neutral as between Pierre's and Paul's we shall simply make both Pierre and Paul, as regarded by us, into fictions. Nor will bringing in acceleration help, for Bergson insists that it too is relative and reciprocal.

It seems to be generally agreed now that Bergson is wrong at least in the letter, though his friends and admirers (Heidsieck, Čapek, Ullmo, to whom my discussion is much indebted) insist he is right in the spirit and anticipated features of relativity theory, just as his 'process philosophy' and treatment of causation and determinism, which we have not yet reached, anticipated the modern physics of matter and quantum mechanics. His opposition to Einstein was unnecessary in terms of his own philosophy.

Einstein, according to Bergson, allows only simultaneity of instants and makes it relative. (See M, 105ff.) But this relative simultaneity, depending on the adjustment of distant clocks, presupposes for Bergson absolute simultaneity. Strictly simultaneity can only be of 'fluxes', i.e. periods, and not of instants, because there are no instants in duration. But we measure duration by spatialising it, imposing instants and therefore simultaneities on it, as explained above (§8). Doing this involves a psychologically basic and therefore absolute grasping of the simultaneity of events with clock-readings and with events in our own inner life. Someone might object that this psychological simultaneity depends on our grasping an objective simultaneity connecting events 'in the same place' and therefore absolute. But this itself is a relative notion, for they would not be in the same place to a microbe, and to insist that our judgements do depend on some ultimate absolute coincidence of place is dogmatically to go beyond observation in a way that contradicts the empiricist spirit of relativity theory itself. (See M, 136 note 2.)

The basic simultaneity for Bergson is evidently that of, say, two sounds experienced as going on together, an experience of harmony. But even if instants do not exist ready-made until we pick them out the simultaneity we experience between them is not of our choosing, and in that sense is real enough, and is what for Bergson lies behind relativistic simultaneities. But the absence of ready-made instants makes the intervals (in the ordinary sense, not Einstein's) between instants relative, thus allowing different rhythms of duration. This is what Čapek expresses by saying that the unique duration Bergson insists on is only topologically and not metrically unique, so that Bergson does not need to contradict Einstein. (See Čapek, 1971: 248ff., 1980: 335, 338.) Similarly for Heidsieck all Bergson needs is that time should go in the same direction in all frames, which relativity allows. Čapek also thinks Bergson confuses the apparent with the unobservable: the slowing of Paul's clock, which is what happens as it appears in Pierre's frame, is something Pierre only calculates rather than observes, and so can be dismissed as somehow unreal. (See Čapek 1971: 243 – 5.)

In the three appendices added to *Durée et Simultanéité* in the second edition, and in an article in 1924, answering criticisms by A. Metz (M, 1432 – 51), Bergson develops his view. He starts, in the first appendix, by comparing the rocket paradox to perspective effects. When Paul walks away he looks to Pierre like a dwarf, but no-one supposes he will really be one when he returns. This is an unfortunate comparison for Bergson. When Paul returns he will no longer *look* like a dwarf, but when he returns in his rocket he *will* mathematically seem relatively younger, i.e. Pierre's mathematical calculations will make Paul out to be relatively younger. This is why it is so important that the real Paul who emerges from the rocket is different from the figmentary Paul of the calculations.

On acceleration he considers an example of a car or train which stops abruptly. (See M, 225 – 9, 1443 – 6.) Surely its occupants will feel a shock which observers outside will not? His way of dealing with this is peculiar, though it fits well enough into his general system. We must distinguish what appears to touch from what appears to sight (M, 228n.). The reason for this is that *feeling* the shock essentially involves the whole body, or at least a whole part of it, and this should for the purposes of physics be regarded as a set of points which when the body is decelerating move in different ways, so that each of them has its own frame of reference. What is illegitimate is to view the situation in more than one of these frames at once. But this is just what perception, and touch in particular, does, so that it gets a result which signifies a real and not merely relative decelerating and which has no meaning for physics, which cannot conflate different reference systems in this way. This splitting of the physical point of view from the perceptual point of view may fit well into Bergson's philosophy, but it seems to have the strange result here that physics must regard a real experience not only as something it cannot say anything about but as something illegitimate or illusory in that it involves looking at the world in different and inconsistent ways at once. When one suffers a shock different parts of one's body move differently, as he rightly points out, and each part can regard the others as moving and itself as at rest. But the fact that they *are* moving differently is an absolute fact, for physics as much as for psychology, and it is on this that one's experience depends.

Later he considers the case where Paul in his car hits a wall behind which Pierre is sitting; Paul dies but Pierre is unharmed (M, 1444 – 5). He rightly points out that the same thing would happen if Paul sat stationary in his car while the wall with Pierre came along and hit him. But this seems to confuse relativity of acceleration (what Bergson is trying to show) with relativity of velocity. In both cases it is Paul who suffers acceleration (described in the first case as deceleration), unless the wall stops immediately it hits him – with unhappy results for Pierre!

Why did Bergson take the line he did? Could he have been over-influenced by his own distinction between intellect and intuition, where intellect arrived for pragmatic reasons at a view of the world that was in some sense improper or illusory? Relativity theory is the work of intellect, and so he may have felt that if its results were paradoxical they should not be wholeheartedly accepted, even though he reckoned to interpret the theory rather than reject it. But why did he think it paradoxical, in view of its affinity to his own philosophy, as explained above, which he certainly did not regard as paradoxical? Perhaps the most important point is that relativity theory belongs to science while Bergson was doing metaphysics. Duration is linked to consciousness. It is one thing to talk of different rhythms of consciousness, as in dreams and waking life. But dreams 'leave not a wrack behind' while Paul's breaking of his twinhood with Pierre is permanent. Duration may allow for different psychological clocks, but relativity lets in different physical clocks, which have nothing to do with consciousness. The weak paradox, like non-Euclidean geometry, is offensive to common sense because its effects are unobservable on the scale of daily life. Whether this has anything to do with Bergson's reactions I do not know, but the distinction is a real one.

III

Free will

1. Introduction. Two forms of determinism

The problem of free will is with us today as much as it was when Bergson wrote, though the threat that loomed large a hundred years ago of an all-embracing deterministic science advancing relentlessly towards its completion has receded. Not only has the massive advance of science spawned embarrassing children like quantum mechanics and the uncertainty principle, but the whole nature of the enterprise has become more open to question. It is by no means as clear as it was that we can talk of the *advance* of science at all, except in the pragmatic or technological sense. I don't seriously want to suggest that science and human knowledge are not advancing. But writers like T.S. Kuhn have at least suggested that the advance is less relentless and unambiguous than we might have hoped or feared. We perhaps think more in terms of a convergence of scientific theories. (See Putnam 1975 – 6.) And no-one, I think, would now talk in terms of science reaching its ultimate completion, even in the more distant future. Of course the contrast I have just drawn between now and then is over-simplified. But it exists, and the threat to free will from a universalistic science was greater then than now. However, this does not mean that the problem has been solved, though now there is more emphasis on asking both determinists and defenders of free will to make clear just what they are maintaining.

Bergson, as we saw, began *Time and Free Will* by describing free will as his main topic, to which the chapters on intensity and duration were leading up. Later duration joins 'voluntary determination' as one of the 'principal objects' of the work, and this gives a more realistic assessment (TF, 226). Bergson is certainly more famous for his treatment of duration than for anything he said about free will, and what he said about duration is far more pervasive in his philosophy as a whole. Yet free will is a topic that should certainly be congenial to one who bases so much of his philosophy on life and consciousness, so what does he tell us about it?

Determinism in the sphere of human action can take two main forms. First there is physical or physiological determinism. Every movement we make with our bodies must obey the laws of nature. We cannot jump over the moon. And the more we study the phenomena of consciousness the more they seem to be correlated with events in the body, and in the brain in particular. Though it seems to make sense to suppose that we could have a thought or sensation without anything relevant going on in the body, science increasingly suggests that we never do; and, even if we could, so long as the laws of physics forbid us to make any bodily movements that have not got adequate physical antecedents such dangling and ineffective thoughts and sensations would have little relation to our mental life as we in fact know it. The only alternative physical determinism has to treating mental phenomena as irrelevant danglers is to treat them as identical with certain physical phenomena. But if the determinism is to be physical, in the sense we are considering, its laws must be expressed in physical terms, and so the mental nature of any physical items that entered into such laws would be irrelevant. The laws would have forms like 'whenever a C-fibre is stimulated, then . . . ', not like 'whenever a pain is felt, then . . . ', so that the pain-like nature of the C-fibre stimulation would be irrelevant.

The other form of determinism is psychological. Mental phenomena may or may not be identical with physical phenomena, but now they are caused by other mental phenomena, which in turn may or may not be identical with physical phenomena, but, if they are, this fact is irrelevant; now it is the C-fibre-stimulation-like nature of pains that is irrelevant. Freudian causation springs to mind here, but more generally we sometimes feel that every action must be caused by a desire or intention or something like that.

It is unlikely that psychological determinism would be held in an exclusive form. We are not disembodied spirits, and we feel no temptation to dismiss the physical as irrelevant in the way that a hard-line physical determinist would dismiss the psychological, as either unreal or irrelevant. Even the identity theory usually says that thoughts and sensations are really brain-states, not that (some) brain-states are really thoughts and sensations. Identity is not always as symmetrical as it seems. An idealist, it is true, would say the physical is unreal and that causal laws could connect only the mental. But in Bergson's day as in our own it has seldom if ever been in an idealist framework that determinism has been discussed, and Bergson himself repudiated idealism. Perhaps even those who accept that some causation is purely mental have felt that causation itself is an idea that belongs primarily with the physical, for it is in the physical sphere that we most naturally see evidence of general laws. Also determinism has a certain appeal, if not to untutored common sense, at any rate to tutored common sense, that idealism lacks.

2. *Bergson's strategy*

Bergson distinguishes these two forms of determinism, and his strategy is to argue first that physical determinism presupposes psychological determinism. This might seem at the outset to contradict what I have just said about the priority of physical determinism for common sense. But Bergson is talking from the point of view of his own philosophy, which does not accept any kind of determinism. He is in fact giving a sort of Humean account of how the determinist comes to hold his position, not claiming that this is how the determinist himself would describe the situation. He thinks therefore that if he can refute psychological determinism physical determinism will fall by the wayside. (See TF, 142 – 3, 148 – 9, 150, 155.)

The real villain is the old doctrine of associationism, which forms a running target for Bergson in various of his works, and which itself relies on faulty conceptions of duration and causality. But the first step in the strategy, linking psychological and physical determinism, seems in fact to have two parts. One part is this: a belief in associationism suggests psychological determinism, but the empirical evidence is too shaky to support a rigorous conclusion, so physical determinism is appealed to because, if it is true, then cerebral events will be determined, and it is at least plausible in simple cases to say that psychological phenomena are correlated with cerebral phenomena, so they too will be determined, and so physical determinism, so reached, is 'nothing but psychological determinism, seeking to verify itself and fix its own outlines by an appeal to the science of nature' (TF, 149). On the face of it this yields not psychological determinism but a special case of physical determinism, the determination of psychological phenomena by physical ones; and this interpretation is supported by the alleged bonus that physical mechanism gains from this process, that it 'would then spread over everything' (TF, 149). But perhaps Bergson means this: the determinist accepts psychological determinism. He also accepts physical determinism, in particular as covering cerebral events. He notices that psychological events are correlated with cerebral ones in simple cases. He generalises these to all cases, and so has two parallel causal series, one physical and one psychological. He then infers that the items in the psychological chain are caused by the corresponding items in the physical chain, whether or not he confusedly goes on thinking of them as also caused by their predecessors in the psychological chain. Finally he then feels that he has strengthened the causal status of the psychological chain by connecting it with the physical in this way, and at the same time has strengthened physical determinism by showing that it covers psychological events, and so can be treated as universal. The sense in which physical presupposes psychological determinism is that a belief in the universal scope of the former comes about in this Humean way through a belief in the latter.

The other part of the first step, linking psychological to physical determinism, argues that no-one would think that physical determinism holds universally unless they believed that the conservation of energy held universally, and they would not think this unless they thought that psychological phenomena were subject to it; and finally this last belief depends on a confusion between time and duration. This confusion takes the form we should expect from our previous discussion: duration is cumulative. It applies (in *Time and Free Will*) only to conscious phenomena, and in them it is lived through and leaves its mark: nothing can ever be the same again. Repetition of a situation in all its detail is a senseless idea, because on its second occurrence a situation would always have at least one feature distinguishing it from what it was on its first occurrence, namely the feature that this *was* its second occurrence – and this would be something qualitative, affecting the nature of the occurrence itself as a conscious phenomenon, not a mere triviality of definition. But time as it appears in physics, and (for *Time and Free Will*) in the inanimate world which physics studies, is a mere parameter for equations, a fourth dimension alongside the spatial ones, and there is no more reason why a time coordinate should not be repeated than why a space one shouldn't. If therefore time and duration are confused one can think of psychological events as essentially repeatable, and so as possible terms in laws; and once we think of them as law-governed we shall be tempted to think of them along with everything else as subject to the conservation of energy, which will then be universal. It is worth noting here – a point I will come back to – that Bergson often applies terms like 'geometrical' or 'mathematical' to this kind of law-governed necessity.

Bergson goes on at this point to say that we must now examine psychological determinism (TF, 155). But has he shown in this last part of the first step that physical determinism presupposes psychological? The assumption he has relied on (i.e. has said the determinist relies on) seems again to be simply that conscious phenomena are determined, not that they are determined in any particular way, psychologically rather than physically. Perhaps Bergson means that we would not assume in the first place that they were determined unless we thought that they were determined psychologically. But, though there is some temptation to treat all actions as determined by desires etc., it is not obvious why this sort of determinism should be thought of as occurring to us first, and when Bergson concludes his discussion of the time/duration confusion by saying that it shows that physical determinism is reducible to a psychological determinism (TF, 155), he certainly speaks as though this latter simply said that conscious phenomena are determined.

3. Bergson on psychological determinism

Turning then to psychological determinism, in its proper sense now, Bergson reaches the second step in his overall strategy. He appeals primarily to associationism.

This is a doctrine with a long and complex history, but what Bergson attacks here is a view like this: any given state of mind consists of discrete elements or features, which can be described in language. As states of mind alter and are replaced these changes can be broken down into replacements of these elements and features, and the process can be catalogued by language throughout. Nothing is ineffable; everything can be grasped, frozen, spotlighted, and described. Suppose I stand up to open a window, but for some reason forget why I stood up before I actually open it. The associationist will say I had two ideas, that of standing up and that of opening the window, and one of them then disappeared, so that I am left with the other. But, Bergson insists, I am not *just* left with the other, or I would presumably sit down again. The idea I am left with has an indefinable qualitative tinge, 'a confused feeling that something remains to be done' (TF, 160), which keeps me on my feet in a state of tension, at least for a bit, until perhaps I recover the other idea. My state of mind while I am standing there is different from what it would have been if I had stood up for some other purpose, e.g. to open the door. Another example: I smell a rose and it brings back memories of my childhood. Again we have not got two separate ideas associated with each other, the smell and the memories, but the memories form part of the experience of the smelling itself; anyone else who smelt the same rose would have a qualitatively different experience. The kind of plurality involved in my mental life is a qualitative heterogeneity, of the kind we discussed earlier. (See TF, 155 – 65.)

The association of ideas raises many problems which we need not discuss. We can agree, I think, with Bergson on two main points. Our mental life is not atomic. It is not made up of elements or features that could occur apart from each other. So far no doubt everyone would agree. To borrow an example from Hume, an atomist if ever there was one, colour and shape cannot occur apart, but we form ideas of them (in his sense) by a 'distinction of reason' (Hume 1888). Are features of our mental life like this? Not, I suspect, if we are concerned with a finite list. My experience in smelling a rose has something in common with yours in smelling the same rose, e.g. sweetness (I ignore the problem of 'transposed qualia'). But it would surely be impossible to list the features of my experience and those of yours and then tick off those in common and those which differ. The second point we might agree on is the ineffability of experience. To take an example where action, like opening a window, is not relevant, consider the experience of knowing that one has just been dreaming but not having the least idea of what the dream was about. I have just described the experience in words, presumably sufficient to enable the reader to pick out relevant cases. But I can go no further. I can invent a word, say 'dreasy'. But this will merely abbreviate 'as of just having been dreaming'. It will do nothing to show why I apply *this* expression, still less to describe how one 'dreasy' experience differs from another.

How far does this concern determinism and causation? Bergson insists that the conservation of energy implies repeatability, and also, as I said earlier, that universal determinism presupposes universal conservation of energy (TF, 152). Actually this last point does not seem obvious. The 'continuous creation' theory of the 1950s presumably rejected the conservation of energy without rejecting the lawlikeness of nature: energy was created in definite quanta in definite circumstances, and all determinism seems to require is that non-conservation should only occur in a systematic and perhaps predictable fashion. But let us assume anyway that determinism implies lawlikeness, which implies repeatability, at least in principle. Total repeatability never occurs, so lawlikeness will involve abstraction. *Time and Free Will*, as we have seen, confines duration to consciousness, but *Matter and Memory* and later works do not, and yet they don't despair of the possibility of doing science in the natural world. Duration is cumulative, and therefore blocks total repeatability; but science only requires the repeatability of abstract features. Different cases where a law applies must be similar in all *relevant* respects, not in *all* respects. (See CE, 226.) Does the cumulative nature of consciousness prevent any kind of repetition? Surely Bergson would not deny that two different experiences of mine might have in common that they were both smellings of a rose? Presumably he would say that the similarities are far more peripheral than in the scientific cases, and that there is no hope of finding laws of the form 'whenever a man smells a rose, then . . . '. It seems plausible to say that there is a large difference of degree between conscious and non-conscious cases here, though it would take some time to spell it out and defend it. But if the difference is one of degree then we cannot rule out causation here in principle.

Some of Bergson's arguments in fact might seem at first to work against his own case, as when he argues that deliberation is often a mere façade, for our mind is already made up (TF, 157 – 8). One might take this as showing that we are not as free as we think we are – that these deliberations are mere rationalisations and not the real causes of our actions. However, this would at most show that sometimes we were less free than we thought, and Bergson's point is to attack associationism: we have here a case where the mind does not work by a linear series of events bringing each other about by the laws of association and culminating in action. An incidental peculiarity of Bergson's argument is that he says the mind goes through this ritual in these cases as a way of pretending to itself that associationism is true after all – as though associationism were not a philosophical theory but a universal common-sense theory about how the mind works: one might wonder why the mind should be so keen to pretend to itself in this way.

4. Nature of causality. Causal and logical necessity

Bergson admits that this argument is not conclusive, in any case, and we should now turn to his treatment of the question itself, whether our actions are deterministically caused. What does he mean by a cause? The opening paragraph of his discussion of psychological determinism has a point of some interest (TF, 155 – 6). He attributes to the determinist a train of thought like this: psychological states are necessitated by preceding states. But 'this cannot be a geometrical necessity, such as that which connects a resultant, for example, with its components', for successive conscious states differ from each other in quality, and so we cannot 'deduce any one of them *a priori* from its predecessors'. So we appeal to experience and try to explain the conscious state of finding a reason for it, which itself will be another conscious state. Bergson agrees that we can always find such a reason, but goes on to doubt whether the relation between the conscious states is a causal one.

This is a significant passage. If conscious states did not, or need not, differ in quality then they could be repeated, as a physical state may recur. In this case the necessity connecting them would be, or could be, 'geometrical'. But what does this mean? In what way *is* a resultant connected with its components? If something moves three miles north and at the same time four miles east it will end up five miles roughly north-east. Its northward movement together with its eastward movement will cause it to be at the new point. But they will not have caused it to move to the new point. Rather its moving thither is what causes it to be there. And its moving north-east is not caused by its moving north and its moving east, but is the geometrical resultant of these movements – it is indeed connected to them by a geometrical necessity, but not by a causal necessity.

Should we conclude from this that Bergson has not learnt Hume's lesson and is simply confusing causal and logical necessity? That would be premature, and he would certainly repudiate the idea. But it is not until nearly fifty pages later that he turns to analysing the notion of causality itself (TF, 201ff.). In the intervening pages he makes two points in particular, as well as saying most of what he has to say about free will. One point is that we could never predict an action with certainty, because we should have to know *all* its antecedents, and if we did we should be in the throes of the action itself; we should be not predicting it but performing it. One might think that this confuses our position at the moment immediately preceding the action ('the last moment of rest', as Aristotle would put it) with our position at subsequent moments. But let us leave this for the present, except to say that Bergson's answer would involve the time/duration distinction. The second point is that, if the determinist says that prediction is a mere epistemological matter and that actions must be subject like everything else to the law 'same cause, same effect', we can reply that the same causes never do recur in the sphere of conscious phenomena because of the cumulative nature of duration. I suggested above that the difference between psychology and physics here is one of degree, and we might anyway ask why it should matter whether the same cause does recur. Surely it would be enough that if it did then the effect would recur? But this appeal to counterfactuals takes us on to Bergson's treatment of causality itself.

5. Determinism as due to confusion of two views of causality

For the empiricists, Bergson tells us, causality is simply regular succession, and as long as this is all there is to it there can be no threat to freedom if, as he has claimed, antecedents in this sphere never are repeated. Since causality is here a purely extensional relation, as we should put it, appeal to counterfactuals is out of order. But common sense refuses to take this Humean view and insists on adding some further element to causality. This can take one of two forms, each of which consists in putting 'an objective connexion of phenomena' in place of 'a subjective association between their ideas' (TF, 204), which is all the Humean view has given us.

In the first form we abstract from the qualitative differences between things, which we attribute to our senses, and construct 'behind the heterogeneity of our sensations, a homogeneous physical universe' (TF, 205). In other words we reduce the external world to primary qualities, but we go further and treat even shape as something given by the senses, 'a concrete and therefore irreducible quality of matter' (ibid.), from which we abstract and reach a purely mathematical formula. As Bergson puts it in his inimitable style: 'Picture then algebraical relations getting entangled in one another, becoming objective by this very entanglement, and producing, by the mere effect of their complexity, concrete, visible and tangible reality' (TF, 205 – 6). Shades of the *Timaeus*, one might think! Bergson, however, appeals not to Plato but to Lord Kelvin, whose theory of vortices in an absolutely homogeneous and incompressible fluid reduces matter to movements. But since these movements take place in a homogeneous medium there will be no detectable difference in the medium at different moments, and the movement is equivalent to 'absolute immobility', since it is not one 'which actually takes place, but only a movement which is pictured mentally: it is a relation between relations' (TF, 206). Real movement involves consciousness, and in space there are only simultaneities. We can see well enough where Bergson is going, if not so clearly how he gets there, and it is no surprise when he compares Descartes's doctrine that the world has to be recreated at every moment. (TF, 208; cf. Descartes, *Meditation* 6, *Principles of Philosophy* I §21.) The successive moments of time are independent, and so there can be no necessary link between them; but science, by abstracting from the differences between them, treats the different moments more and more as though they were the same. If they really were the same the link between them would be one of identity, and causality, as applied by science, is a relation which indefinitely approaches identity, and so is thought of as a logical relation between cause and effect (TF, 208). But the more we assimilate causality to identity in this way, the more, that is, we treat it as a necessary relation, the more we treat things in the world as lacking duration and so as different from ourselves. We ought therefore to realise that causality does not apply to our own actions, and so that we have free will. (TF, 210; cf. 215.)

Incidentally Bergson refers again to Lord Kelvin at MM, 265 – 6. Here he uses the vortex theory to support his view that the discontinuities which our senses impart to the world – the arrangement of the world into discrete objects which our mind engages in for practical reasons, as we will see later – do not belong to the world itself, which is a world of change, not of things. We saw in the last chapter that *Matter and Memory* differs from *Time and Free Will* by allowing duration to things themselves, and we might express the relation between the two passages like this: in *Time and Free Will* science takes the common sense notion of causation to the limit, where it becomes the logical relation of identity, and in doing so science treats the world as lacking duration, which indeed it does. In the world itself there is neither duration nor necessary connexion; but science puts necessary connexion into the world full-bloodedly so that it becomes identity, while common sense does so half-heartedly, so that in a confused way it both becomes necessary and spans time. He talks of our 'feeling', and talking in such a way that we 'assert thereby', that things do not endure (TF, 209 – 10). This may sound as if he is talking dialectically, but his summing up makes clear that he is giving his own view at this point (TF, 227). In *Matter and Memory* science again spatialises the world and treats it as lacking duration, because of the practical advantages of doing so; any intellectual as against intuitive approach must do likewise. But science also helps us to see that the discontinuities foisted on the world for practical purposes by perception do not really belong to it. For it is because of this that science takes the form it does, as illustrated by the vortex hypothesis. The world now does have duration, so science falsifies it, albeit inevitably. But the world still has no necessary connexions linking things across time. Necessity is still to be analysed as the logical necessity of identity, which science uses as it did in *Time and Free Will*. In *Time and Free Will* the world lacks discontinuities (though this is not emphasised yet), lacks necessary connexions except identity, lacks duration, and is therefore spatialised, as science describes it. Common sense adds the discontinuities and confusedly applies to them the necessary relation of identity, now regarded as spanning time. In *Matter and Memory* the world lacks discontinuities, lacks necessary connexions except identity, but has duration and is therefore not spatialised, and so to that extent science falsifies it. Common sense behaves as before. However, whether Bergson would agree with this way of comparing the passages I do not know. In the autobiographical note which he wrote for *The Creative Mind* of 1934 he refers to both passages and says simply that the second tries to get nearer to what he wanted to say (CM, 84 (71 – 2)). But he adds that even in *Time and Free Will* he thought of reality in itself as moving and changing, immovability and invariability being only 'views taken' of it. He does not mention duration here, but perhaps we should take his early denial of duration to the real world as simply an unstable position *en route* to his main view. (See also chapter 2 §11 above.)

In the second form taken by the further element common sense adds to purely extensional causality we project on to nature the experience we have in ourselves when a later conscious state, though not contained in an earlier one, is prefigured in it as something we can achieve by effort (TF, 211 – 15). We think of events in nature as connected in this sort of way, and in fact as flowing into each other as our conscious states do, and therefore as having duration. But since effort is not always successful the conscious states are clearly not identical; and so, on this model, the states of the world will not be identical, and the later state will not be inevitable given the former. Human actions will not be inevitable then since even natural phenomena are not, and so free will is preservable.

Both these ways of regarding causation then can preserve free will, but trouble arises because we insist on fusing them together. We think of events in the world as separate and distinct because of the second view, yet linked by the identity relation because of the first view, and so as necessarily connected despite being distinct. We then think of conscious states as being on the same level as events in the world because of the second view, and so think of them too as distinct but necessarily connected, so that there is no free will. (See TF, 215ff.)

6. Causal and logical necessity again

In the above passages Bergson has offered an account of how we adopt universal determinism by conflating two approaches to causation. In 1900 he gave a lecture entitled 'Note on the psychological origins of our belief in the law of causality' (M, 419 – 35), but which is really on how we acquire the concept of causality rather than how we come to accept determinism. He first dismisses three possible accounts, pointing out rightly enough that the amount of regularity that we actually observe is often wildly exaggerated. He then gives an account of how the passive data of vision are normally followed by tactile data as the baby, or higher animal for that matter, probes the object with its limbs, with the result that 'the habit is created of expecting these tactile impressions when these visual forms appear' (M, 424); a corresponding account holds of those born blind (M, 427). But Bergson goes beyond Hume in insisting on the active nature of the actual probing involved, and it is this which we transfer to the outer world (M, 424 – 5).

This raises a certain problem of interpretation. The point seems to be that we think of one event in the world as 'producing' or 'dynamically causing', not just being followed by, another just as our visual experiences 'produce' our tactile ones through some action on our part. We then call such production necessary because of the 'invariable correspondence between the visual impressions and the tactile impressions' (M, 425). Perhaps 'invariability' is more justified here than with the spurious regularities he dismissed in his preliminary remarks. But he also dismissed there the theory, which he attributes to Maine de Biran, that we get our idea of causation from our own force of action (M, 421 – 2). This he dismisses because we think of our own actions as free, and so cannot draw from them the idea of causal necessity. But surely we might if we think of our actions as necessitat*ing*, not as necessitat*ed*, which is not too far from what Bergson is saying himself.

But to return to Bergson's relations with Hume: some time ago I asked whether Bergson had learnt Hume's lesson or was still confusing logical and causal necessity. Obviously he does not himself think that events can be both causally and logically related. The question rather is whether he thinks, and is justified in thinking, that anyone who claims that two events are causally or necessarily related is committed to saying that they are logically related.

Hume himself gives a causal account of how we acquire the idea of causal necessity. There is no vicious circle in his doing this, so long as the account he gives claims no more than to be causal in his own sense of 'causal', for which purpose it need involve no more than regular sequence. As to whether there is any stronger sense of 'causal' than this, he could consistently allow its theoretical possibility, while disclaiming any knowledge of it himself. (See Hume 1888: 168.) He has a relatively clear and unambiguous account of what someone is committed to who says that two events are causally related: that the relevant regular sequences exist.

Similarly Bergson can consistently give a purely extensional account of how the mind comes to its idea of causal necessity – provided he allows that regular sequences are to be found in the mental sphere, which is just what he denies, at any rate where action is involved. He does allow that we are not free to have what sensations we like when our senses are physically stimulated, and that such sensations and 'many other psychic states' are 'bound up with certain determining conditions' (TF, 146). Whether these other states include those involved in getting the idea of causal necessity is not obvious. But, if we allow that this thin causal account can be given, it is still not clear that a richer account could, with either of the two richer ideas of causality that Bergson attributes to common sense, as we have just been seeing (TF, 203ff.). Since the acquisition occurs over time it cannot involve logical necessity, and if it is assimilated to action it can hardly be something guaranteed to occur regularly. So Bergson can have his causal account provided it remains a thin one; he can be as well off as Hume here, provided he is equally satisfied with a thin account. He makes no attempt, I think, to deal with problems like that of distinguishing causes from coincidences. In a letter written in 1903 he says, 'In my view determinism is perhaps radical in brute matter' (M, 585). Whether this refers to anything beyond the thin sense is not clear.

The question then comes to this: has Bergson any concept of causation other than the thin one (regular succession) which he thinks anyway does not apply in the sphere of life and consciousness? We will come later (§8) to the 'psychological causality' he thinks does apply there. But here his point may be that anyone who claims to have more than a thin concept must confusedly think that logical necessity can connect events at different times. Why should Bergson think this?

Let us return for a moment to the example of the parallelogram of forces. Is it a logical fact or an empirical one that something which moves north and moves east will have moved north-east? Surely a logical one. Presumably then it is also logically true that if it is caused to move north and caused to move east it will have been caused to move north-east. Yes, but only if it *is* caused to move north and caused to move east. It might have been that any force that causes something to move north only does so if nothing else causes it to move east at the same time. It might have been that anything hit simultaneously by a force that would normally move it north and one that would normally move it east moved south-west, or exploded. As a principle of dynamics the parallelogram of forces is empirical. Any necessity it has is not logical. But confusion is possible both for the reason just given and because its necessity might eventually be analysed in terms of logical deduction from or in accordance with other principles accepted independently.

This last possibility is one I think Bergson never considers. He seems to assume that one event can only be necessarily linked with another if the occurrence of the one entails that of the other. If it does, then indeed they cannot occur at different times. But, he thinks, if we look for the event which is the immediate cause of another we are led to shorten the time-gap between them more and more, ending up with the 'd*t*' of differential equations, which is not a finite interval at all. We are dealing with tendencies, but 'always speaking of a given moment . . . not of flowing time', and so we are dealing with 'a world that dies and is reborn at every instant', which Descartes required God ever to recreate (CE, 22 – 4). In a rather similar passage later Bergson discusses induction (CE, 225 – 8). He takes a set of circumstances known to resemble a previous set in all relevant respects but one, where we induce the presence of this one in the later set. The induction would be certain only if we could superimpose the one set on the other like two geometrically similar figures, where we deduce the final remaining correspondence. Induction in other words becomes certain just so far as it becomes deduction. But the difference in time prevents this superimposition, which can be asymptotically approached in imagination but never finally reached except by reducing the time difference to zero, so that the situations cease to be two.

It seems in fact to be this idea that lies behind Bergson's claim, which we discussed earlier, that for science time could be speeded up indefinitely, so that in the limit everything happened at once. If time is not a kind of force why do causes have to wait for their effects to occur? (CE, 358) One might say, because otherwise things would have inconsistent predicates. But that does not explain why things have the temporal intervals they have. There can be no answer in terms of 'abstract' time for Bergson; only the existence of real duration can account for the delay. The future is not determined by the present but is being constantly created (CE, 358 – 9). How this fits the 1903 letter quoted on p.79 is unclear. Could it be that his slow-melting sugar did not count as 'brute matter'? But what about inorganic chemistry? (Cf. again chapter 2 §11 above.)

A way of expressing one thought that underlies all this is a sentence from Čapek: 'It is clear that one must choose one of two assertions: *either* real succession with an element of real contingency, *or* complete determinism with the total absence not only of possibilities, but of succession as well' (Čapek 1971: 111). (Cf. CE, 41, 7.) This seems to exclude the possibility that something not now real should be inevitable because the cause, i.e. the sufficient conditions, of its future existence already exist. The thought is evidently that the sufficient conditions for something cannot exist unless it does, and furthermore cannot exist now unless it exists now. Both these things are true for logically sufficient conditions, and the former is true for causally sufficient ones (they would not be causally sufficient unless the effect existed sometime). But why should the latter be true for causally sufficient conditions? If it is not true, then indeed we shall have a puzzling relation of inevitability to account for. But to call it puzzling is not to call it impossible. Its existence might be confirmed by as many predictions as we cared to make. Čapek says determinism cannot account for the posteriority of an event (Čapek 1971: 109). But perhaps the sufficient conditions just *are* sufficient for it to occur at a certain later time. Again we have a puzzle but not yet a contradiction. Čapek adds that if duration is real the future is unreal and its coming to be is a genuine and not just an apparent novelty (Čapek 1971: 108). But when is a novelty genuine? Does Čapek distinguish between calling the future unreal because still future and calling it unreal because uncaused or not inevitable? The past is inevitable, but does not exist now. It is real only in the sense of not being fictitious: it really did occur.

7. Libertarianism

We have now discussed Bergson's attitudes to determinism and causality. But what does he say about libertarianism? Some of what he says is surprising, at least for anyone who expects him to be an orthodox libertarian. He repudiates libertarianism, along with determinism, and for much the same reason: they both replace the reality with the symbol, in the sense in which the sciences replace concrete duration by abstract time (TF, 180 – 1). The libertarian insists that whatever we do, when exercising freedom, we could have done otherwise even though all the circumstances up to the time of acting were the same. This has always been a sticking point for determinists. If *everything* were the same, including all our motives etc., and yet we had chosen differently, what could our choice be but what T.H. Green called 'an arbitrary freak of some unaccountable power of unmotived willing' (Green 1883: §110), and how could that possibly be what we demand for responsibility? (See Hobart 1934 for a development of this compatibilist attack.) Perhaps it was this sort of consideration that Bergson had in mind in hastening to distance himself from the libertarians. He certainly thought that some third way is needed, and he reduced the determinist and the libertarian alike to uttering tautologies: the determinist says, 'The act, once performed, is performed', while the libertarian says, 'The act, before being performed, was not yet performed' (TF, 182). But, though we may sympathise with this sort of motive for rejecting libertarianism, the argument Bergson actually uses to do the rejecting looks distinctly suspicious. He imagines a self which goes through a set of experiences until it reaches a point, O, at which it regards itself as faced with two possible directions to continue in, OX and OY, between which it must choose. Actually, he thinks, X and Y will not remain constant, for states of the self are not repeatable, as we have seen, and so the envisaged courses of action will be constantly changing, at least in the sense of constantly appearing to the chooser in a different light as he accumulates the experience of considering them. But common sense likes cut-and-dried alternatives, and so tends to regard OX and OY as two constant alternatives facing it:

> we thus get an impartially active ego hesitating between two inert and, as it were, solidified courses of action. Now, if it decides in favour of OX, the line OY will nevertheless remain . . . *waiting in case the self retraces its steps in order to make use of it.* (TF, 177, my emphasis)

For Bergson the error lies in separating the self from the path which it follows; for if we do, then, since we know that OX was in fact followed, there will be nothing to stop the determinist saying that the state of the self at O already implicitly contained a tendency to go along OX (TF, 178). Bergson's own position is to deny that there is any point O or paths OX and OY. To ask whether the self, having arrived at O and chosen OX, could have chosen OY is therefore meaningless, for to ask it is 'to admit the possibility of adequately representing time by space and a succession by a simultaneity. It is to ascribe to the figure we have traced the value of a description, and not merely a symbol' (TF, 180). A few lines later he complains that the figure represents a 'thing', not a 'progress'.

I called this argument suspicious because of the clause I italicised. It makes the libertarian hold that OY, the rejected course, will still somehow be there and can be returned to. In considering the choice in retrospect we will probably imaginatively return to O and try to reconstruct the scene as it was. But there is no guarantee that we can do this, even imaginatively, for to repeat the choice would be to repeat it with our present self, wiser from experience. To ask if we *would* choose OY if returned to the scene is ambiguous. If our present self were presented with the choice it might well choose OY, but if we were returned to O without benefit of the intervening experience, presumably we would choose OX, unless choice is a purely random matter, since OX is what we did choose. But what has that to do with whether we *could* have chosen OY? The libertarian's whole case is that 'could not have chosen OY' does not follow from 'would not have chosen OY', any more than it follows from 'did not choose OY', because, in a *complete* repetition of the situation, it is hard to see how 'would not have chosen' can differ from 'did not choose', except purely arbitrarily. True, the libertarian has not yet given us an account of 'could have chosen otherwise'; but neither has he been refuted. It is tempting to think that it is the 'arbitrary freak of unmotived willing' view that is Bergson's real target here, and that his own view, which we shall now come to, is indeed libertarianism, as opposed to a travesty of it. But what self-respecting libertarian would ever dismiss as meaningless the idea that we could have chosen otherwise than as we did?

8. Bergson's own view of free will

Bergson's own view is that a free act is one where 'the self alone will have been the author of it' and 'it will express the whole of the self' (TF, 165 – 6); 'we are free when our acts spring from our whole personality, when they express it, when they have that indefinable resemblance to it which one sometimes finds between the artist and his work' (TF, 172). 'Freedom is the relation of the concrete self to the act which it performs' (TF, 219).

This view has various corollaries. For one thing freedom is a matter of degree, and one of Bergson's criticisms of libertarianism is for making freedom 'absolute' (TF, 166). Free acts – presumably he means fully free ones – are exceptional. Our ordinary run-of-the-mill habitual actions may well be determined.

> It is at the great and solemn crisis, decisive of our reputation with others, and yet more with ourselves, that we choose in defiance of what is conventionally called a motive, and this absence of any tangible reason is the more striking the deeper our freedom goes. (TF, 170)

Many years later, in 1916, Bergson said much the same in a lecture in Madrid: 'The will is precisely this: the faculty of introducing something absolutely new, apparently, in the world' (M, 1203: my translation of a French translation of a Spanish translation of a journalist's shorthand notes of the lecture).

This emphasis on acting without motive or reason creates a certain tension with the view of freedom as expression of the self, which suggests, if anything, determinism by character, a danger Bergson foresees and tries to defuse by saying, 'Our character is still ourselves' (TF, 172). But acts that fully reflect one's character are not acts done without motive or reason – unless one is a flighty sort of person. Perhaps he means, without a determining motive or reason. But that leaves open the whole issue: how *are* acts related to their motives or reasons? In the 1903 letter, referred to on p.79, he insists that freedom is not mere absence of physical determinism, though it does indeed involve 'zones of indetermination' where life exists, and that freedom is nothing if not 'psychological causality itself'; but this kind of causality is different from physical causality in that it does not involve an 'equivalence' between an act and its antecedents, but is essentially creative. He evidently intends to distinguish it in this way also from the kind of psychological causation I mentioned at the beginning of this chapter. His correspondent (L. Brunschvicg) had objected that 'we feel ourselves the freer the clearer knowledge we have of the reasons which have determined us' (M, 586). Yes, replies Bergson, but these reasons only determine us at the moment when the act is virtually accomplished, and the creativeness comes in in the process whereby these reasons *become* determinant. There may be reasons for this process itself, but they cannot be of the same sort, on pain of a regress. Really the process is that of the personality as a unified whole, or rather the reasons, and the reasons which make them effective, and all our psychological states, both ideas and feelings, are so many external views of this moving unity which is the person – views taken by consciousness when it *analyses*. In particular he rejects any contrast of will and feeling as the sphere of liberty against intellect as the sphere of determinism.

I have summarised this letter at some length since it brings out quite well the tension I referred to, though how far it resolves it is less clear. Certainly Bergson is right in saying we must avoid a regress of reasons for reasons – though, if someone is affected by a reason which someone else, equally conscious of it, is not affected by, we feel inclined to ask for *the* reason for this, even if we cannot ask for *his* reason for it. Bergson would presumably say simply, 'That's the sort of chap he is', and perhaps that is all that can be said, though there is still a certain tension between his being *consistently* that sort of chap and his displaying novelty and spontaneity. It is the relation between authenticity and consistency that is the problem, though it is certainly not a problem only for Bergson; but he may exacerbate it by his insistence on novelty and unpredictability.

There is a certain affinity between Bergson's view of free acts as those representing the whole person and a view like that of R.M. Chisholm who distinguishes between transeunt causation of events by events, and immanent causation, or the causation of events by a continuant, the self (Chisholm 1966). There is the same difficulty about saying what the relation is between the self and the event it causes, and also a difficulty about the time of the event; why did it happen when it did? As for the former difficulty, Bergson makes clear at the end of chapter 3 of *Time and Free Will*, after he has called freedom 'the relation of the concrete self to the act which it performs' (TF, 219), that this relation is indefinable, for it involves duration, and we cannot analyse duration without transforming it into extensity; only things, not processes, can be analysed. We are back with the old sin of trying to represent time flowing, and not just time flowed, by space, of substituting the trajectory for the motion. (Cf. TF, 221.)

Bergson continued to show signs of tension if not uncertainty in his views in this area. In a discussion note of 1910 he seeks a notion between 'moral liberty', which involves complete independence of action from anything but the self, but suggests a *necessary* dependence on the self, and '*liberum arbitrium*', which implies the equal possibility of contrary alternatives (M, 833 – 4). The notion he seeks is between these two, but if anything, nearer to the latter. He also does not want to confine freedom to the moral, which brings us back to the question of the scope of free action. In 1918, in a comment on a written version of a conversation he had held, he repeated that only important or decisive acts could be free, adding that they were not absurd or anti-rational. To confine free actions to important ones seems better than confining them to moral actions, as has sometimes been done, and is rendered easier by letting freedom have degrees. But it still sits uneasily with our intuitions. Surely I am just as free to choose tea or coffee for breakfast as in any other choices I make, and indeed it is in the more portentous spheres that I am likely to feel that my decision really owed more to parental influence or Freudian experiences or unconscious desires than seems obvious at first. But Bergson himself is not unambiguous in what he says about this. In a none too clear discussion right at the end of *Time and Free Will* he seems to think that our freedom is wider than it might seem when we reflect intellectually on it, but that this reflecting may itself, if we are not careful, inhibit or limit our freedom: 'the very mechanism by which we only meant at first to explain our conduct will end by also controlling it' (TF, 237). Renouvier seems to have thought of habitual voluntary actions as like reflex actions, and to have 'restricted freedom to moments of crisis. But he does not seem to have noticed that the process of our free activity goes on, as it were, unknown to ourselves, in the obscure depths of our consciousness at every moment of duration' (TF, 237n.). Bergson now seems to be saying rather that all action that occurs within duration, and so all action, is free, which on his own terms it ought to be, since it is not repeatable, and so not subject to laws or predictions, as we saw earlier.

In 1911 Bergson gave a double lecture in Oxford on 'The perception of change' (included in *The Creative Mind*). In the penultimate paragraph of this he says that discussions of free will

> would come to an end if we saw ourselves where we are really, in a concrete duration where the idea of necessary determination loses all significance, since in it the past becomes identical with the present and continuously creates with it – if only by the fact of being added to it – something absolutely new. (CM, 184-5 (156))

The same book begins with a double introduction in the form of an intellectual autobiography, in the course of which he refers to the problem of how there could be free will in a world where physical determinism reigned as one he had avoided in *Time and Free Will* but finally came to grips with in *Matter and Memory* (CM, 86 (73)). We saw earlier that in 1903 he posited 'zones of indetermination' where life existed. But it is in *Matter and*

Memory that he devoted his main attention to the relations between mind and matter and to related questions, and it is some of these that must occupy us next.

IV

The metaphysics of change and substance

1. Introduction

I said at the outset that Bergson was sometimes called an idealist, but emphatically repudiated any such label. He also rejected realism, at least of a materialistic variety, despite adopting it in the passage I referred to 'if one must choose'. The new 'Introduction' which he wrote for *Matter and Memory* in 1911 opens by calling the book 'frankly dualistic' because it 'affirms the reality of spirit and the reality of matter'. This is something he does insist on, but his attitude to dualism is itself ambiguous. He rejected 'ordinary' or 'vulgar' dualism ('dualisme vulgaire': MM, 294, etc.), and attacked parallelism in a lecture/discussion of 1901 (M, 463ff.)

Ordinary dualism is too sharp. It postulates two systems, but cannot explain why there are two, and in one form (subjective idealism) it tries to derive one of these systems from the other, the world of science from the world of consciousness, while in its other form (materialistic realism) it does the opposite (MM, 14 – 16). We might think it odd to call either of these systems dualistic and indeed Bergson does not explicitly do so, and distinguishes 'materialists' and 'dualists'; but he talks of what they are 'fundamentally agreed on' (MM, 11), and does seem to think of these various systems as ultimately amounting to dualism. The point seems to be that idealism is forced to say something about the world of matter – it cannot just ignore it – and so it tries to derive it from the privileged world of consciousness, while materialism does the opposite for the same reason. Both of them in fact use a distinction based on space where they should use one based on time, for 'matter is considered as essentially divisible and every state of the soul as rigorously inextensive' (MM, 293). (See

MM, 293 – 5.) No doubt there is some truth in the idea that each of idealism and materialism has some difficulty in explaining the phenomena that seem to ground the other, and has to resort to talk of (mere) appearances, without explaining why the appearances should be as they are. So does Bergson do any better?

2. *The escape from realism and idealism*

Matter and Memory is the hardest of Bergson's works, and one of its unnerving features is the casual way in which 'images' are introduced in its second sentence; by the time of the new 'Introduction' they are at least clad in inverted commas. (Cf. CM, 90 (77).) Images are important, for they are the nearest Bergson comes to providing an ontological ground floor for his system; but what are they? Certainly their being does not consist in being perceived. They are 'perceived when my senses are opened to them, unperceived when they are closed' (MM, 1), and they interact according to laws (the laws of nature), or at any rate they appear to do so. In the new 'Introduction' they are introduced more formally: 'by "image" we mean a certain existence which is more than that which the idealist calls a *representation*, but less than that which the realist calls a *thing* – an existence placed half-way between the "thing" and the "representation"' (MM, xi-xii). Matter is an aggregate of these images (MM, xi), but the term 'image' is itself a concession to idealism, signifying that 'every reality has a kinship, an analogy, in short a relation with consciousness' (MM, 304 – 5).

Bergson's position is avowedly a compromise between realism and idealism, at least so far as matter is concerned (MM, xi), and it also claims to be the view of common sense: ordinary objects do not for common sense exist only in or for the mind, but they do include secondary qualities; 'the object is, in itself, pictorial, as we perceive it: image it is, but a self-existing image' (MM, xii). This sounds like naive realism, and is partly presented as such: 'We place ourselves at the point of view of a mind unaware of the disputes between philosophers' (ibid.). But the trouble with naive realism is that it is naive. Just because it is innocent of philosophy it is likely to be inconsistent or in some other way inadequate. Why otherwise should philosophy exist? We might start from a naive realist position, hoping that it will turn out to be justified, but the fact that we have started from there is not itself a justification, and Bergson's position, as it develops, hardly continues for very long to be that of common sense. It is all very well to say the object exists 'as we perceive it', but as who perceives it? If I perceive the wind as cold and you perceive it as hot, which is it? Common sense is not perturbed by this: one of us is mistaken. But it is here that naive realism starts to dissolve, and philosophy takes over. As soon as we consider such difficulties we are no longer naive. But Bergson, though well enough aware of 'the disputes between philosophers', some of which he goes on to discuss, does not raise this particular problem, I think. He seems to regard common sense as not only methodologically justifying a starting-point but metaphysically justifying a position.

The trouble is not that the image must coincide with what is perceived. As we saw above, images can exist unperceived. There is no difficulty then about the wind being hot without my feeling it to be so. But I feel it to be cold. Was it therefore both hot and cold? The obvious way out would be to say there are two winds, one cold and one hot, and we perceived one each. But this would seem arbitrary, for the two winds would have so much else in common, such as their locations and their effects on surrounding leaves; should we not end up with having two winds in the same place?

The question how images are individuated is not one that Bergson shows much interest in. We have seen that matter – matter in general – is an aggregate of images, and he often refers to the body as *an* image, but the use of the indefinite article seems casual, and he shows little interest in deciding where one image ends and another begins. This might not seem to matter much; for where does one object end and another begin? But it is questions like this one about inconsistent predicates that show that we must indeed be told something more about what counts as an image, even though it may not matter if images overlap in the way objects do when an object's parts are called objects.

One thing Bergson wants to avoid is conceiving of matter as an unknowable Lockean substrate (at least on the traditional interpretation of Locke) – a substance 'of which . . . we have no image' (MM, 9). This apparently flagrant reversion to the ordinary notion of images, as something we 'have' and which are 'of' other things, is deliberate, I think. He is not just confusing two senses of 'image', nor two senses of 'of'. (Cf. 'image of Vesuvius', 'image of a unicorn' for this possibility.) His point seems to be rather that if matter were something completely cut off from our experiences it would be unintelligible how it could ever give rise to those experiences. Speaking of 'many philosophers', i.e. 'ordinary dualists', though he does not call them so here, he says they 'invent an incomprehensible action of this formless matter upon this matterless thought', the 'matterless thought' being a 'consciousness empty of images' which they postulate as something over against matter and in which 'representation will issue as by miracle' (MM, 9).

In the next few pages Bergson develops an argument which appears to beg the question in favour of his own view of objects as being images (MM, 9 – 12). His opponents, he thinks, are committed to treating the brain as something existing in isolation from its environment. They show this by trying to generate our picture of the world as a whole from the brain, i.e. making the brain responsible for our having the conscious experience of the world that we do have. But the brain cannot do this because it is itself an image, as are its 'cerebral vibrations', and: 'Images themselves, they cannot create images' (MM, 10). The begging of the question comes in this assumption that the brain and its 'vibrations' *are* themselves images. But perhaps we can offer a more sympathetic interpretation. The philosophers Bergson has in mind are evidently impressed by the idea that our entire conscious life, including all our knowledge of the world, depends on and is therefore somehow generated by our brains. But, he replies, it is all very well to say the brain only mediates this knowledge, which is generated by things outside, but how are we supposed to know this or know of these things, if we have no access to them except via our brains? (I will return to the fallacy in this step near the end of chapter 5.) Yet we must know of them if we are to state the theory at all, since the brain itself is one among such objects – unless we deny or ignore this and treat the brain as existing in isolation and accessible to us by its intimate relation to us in a way that ordinary external objects could not be. To use a famous simile, the brain is a telephone exchange which gives us messages about a world we have no other access to, but somehow we manage to understand the messages. Bergson himself uses the telephone image a little later, but in a different way (MM, 19). The question-begging element comes in, on this interpretation, because Bergson simply states the problem in terms which presuppose his own solution to it, as though the problem were: 'How can an image generate images?' But his underlying point seems to amount to this: How can we coherently say that all we know of the external world is through images of it generated by the brain, since the brain itself is part of that world, and so it itself can only be known by images, so how can we know there is anything at all except images? Yet we claim that brains and other objects are not images. If on the other hand we claim a special privileged knowledge of the brain this itself would already give us knowledge of the world, of which our brain is part, without our having to wait for this brain to generate images.

Our own solution, I take it, would be to say that indeed we do not know of things solely through images without knowing the things themselves. We do know the things themselves, whatever account we give of how we do so. In any case Bergson is surely right in opposing any view which says we acquire knowledge of the outer world solely on the basis of knowing something entirely different. Later he describes the view (or a view) he is objecting to as trying to construct the extended world out of unextended sensations (MM, 44, 47), and rightly asks 'how could these sensations ever acquire extension, and whence should I get the notion of exteriority?' (MM, 43). Still later we find that both realism and idealism treat perceptions as 'veridical hallucinations', 'states of the subject, projected outside himself' (MM, 73). (Cf. MM, 318.) Realism and idealism 'differ merely in this: that in the one these states constitute reality, in the other they are sent forth to unite with it' (MM, 73). Whatever the precise sense of these rather opaque final words the point seems clear enough: we must not put impassable barriers between the world and our experiences of it.

Bergson often talks of his own position as an intermediate one, as we have seen. It is between those of Descartes and Berkeley, at least in the sense that it wants to 'leave matter half way between the place to which Descartes had driven it and that to which Berkeley drew it back – to leave it, in fact, where it is seen by common sense' (MM, xiv). Realism makes perception an accident, idealism makes the success of science an accident (MM, 16). We have seen that he rejects 'ordinary' dualism, though he ends up as a dualist. He seems sometimes to hold a view reminiscent of neutral monism, especially in the version of William James. (See MacWilliam 1928: 125 – 7.) Instead of 'neutral stuff' we find in James 'experiences', suggesting a certain bias towards the mental in its 'neutrality', at least so far as choice of terms goes (James 1912). The term 'image' suggests a somewhat similar bias. (See especially MM, 13.) An affinity with James would hardly be surprising in view of their mutual admiration and the strong pragmatist element in Bergson, of which more anon. But the fact remains that he regarded himself as a dualist, and nowhere, I think, expresses any explicit adherence to neutral monism as such.

3. The role of images

What then are images and what is their role? An image can be without being perceived (MM, 27), but the difference between being and being consciously perceived is one of degree, not of kind (MM, 30). An image can be 'present without being represented' (MM, 27), i.e. without being perceived. But Bergson's main point is that where it *is* represented this involves subtraction, not addition. The doctrine as a whole involves material we have not yet come to, concerning perception and memory. We have seen something of its motivation, in the desire to avoid the unbridgeable gap between mind and matter that arises on 'ordinary' dualism, a gap that is epistemological in so far as we cannot get from our sensations to the outer world and metaphysical in so far as the Cartesian problem of interaction arises. Perhaps we can see something of the doctrine itself by way of an analogy, though the analogy I am about to present is certainly not one used by Bergson himself; he would probably have called it misleading, and so it is if pushed beyond its purpose; but I introduce it because it lets us start from something nearer to the concepts we are most of us familiar with. Bergson himself tends to take his own frame of thought for granted, as I said above in connexion with his use of images, and he seems to underestimate the problem of communication he faces.

Imagine a vast concourse of atoms randomly jostling in the void. Each atom reacts indifferently to all it comes in contact with; it is affected by all of them equally, at least in so far as it happens to meet them, and its own movements are a function entirely of what is going on around it. In a sense these other movements mean nothing to it; it reflects them all equally. But now suppose some atom or group of atoms starts acting spontaneously. Its movements are no longer entirely a function of those surrounding it, but it cannot ignore these, since other things being equal it will still be pushed around by what hits it. But since it can initiate movements of its own it can steer a course so as to avoid certain impacts and engineer certain others. It will no longer reflect all the movements around it indifferently, but will reflect some of them, differentially. It will reflect only a selection of all these movements, fewer than originally reflected, but those it does reflect will 'mean something to' it.

To cash our metaphor we now replace atoms by Bergsonian images. The images 'are', out there, and interact with each other indifferently, whatever this may amount to. Then one of them (a 'body') becomes a centre of life and starts acting spontaneously. In doing so it reflects the other images differentially, and this amounts to perceiving them, for perception is always partial and from a point of view. The other images, which all 'are' in so far as they interact, are now perceived in so far as they enter into these differential interactions. Their being perceived is something less than their merely being, because it involves only some of the interactions they are engaged in, namely those which depend on the perspective, and also on the interests, of the perceiver. At the same time this representation of the material universe is not something that emerges from us; rather we emerge from it, in the sense that each of us is one of these centres of spontaneity that emerge in the flux of images. (Cf. MM, 54, and also MM, 26 – 30.)

Obviously this still leaves us in a good deal of obscurity, much of which cannot be resolved anyway until we have seen more about the nature of perception and memory. In particular I have been emphasising the monistic rather than dualistic features of the system, and Bergson himself shows some signs of realising that his readers might sometimes lose sight of his dualism. (See MM, 294.) It is not till chapter 4 that he draws out the implications of his views for the relations between mind and body, but to anticipate a little we can note that one thing he does insist on is that the difference between the extended and the unextended is one of degree and that in general the distinction between spirit and matter must be made in terms of duration rather than of space, and that this distinction will be a matter of degree. (See MM, 294, 295 – 7.)

But let us return to the nature of images, and their relations to objects. We find that 'the material world is made up of objects, or, if you prefer it, of images, of which all the parts act and react upon each other by movements' (MM, 74). This cheerful indifference between objects and images is in line with passages like those where the brain is called an image, and the present passage suggests that images are the building blocks of the world, rather like bricks garnished with a psychological perspective.

But is this really Bergson's view? One of the most familiar ways of classifying him is as a philosopher of change or process, in a tradition including Hegel and Whitehead and stemming from Heraclitus. Actually he repudiated the connexion with Heraclitus for reasons we will come to, though in 1913 he wrote in a letter a passage very Heraclitean in spirit: 'Reality flows; it is already far away from the word which thought to hold it; and it is found in our formulas to just that extent to which the current of the river is found in the water we draw from it' (M, 989).

4. Categories. Change and essence

The mainstream of Western metaphysics draws heavily on Aristotle's doctrine of categories, whose keystone is the primacy of substance. There is no category of change as such, but the two categories of action and passion can be said to cover it. What is important is that there can be no change unless something changes; one cannot, as it were, have changes floating around on their own. The same applies to all the other categories. One cannot have qualities without things qualified, quantities without quanta. Aristotle has an ontology based on 'primary substances', men and horses, chairs and tables.

How far does Bergson deny this, and how far, if he does, is he justified in doing so? Čapek thinks he does deny it and is justified. (See Čapek 1971 part III chapter 10, on 'change without vehicle and container', and chapter 15, on Strawson's auditory worlds.) One of Bergson's most definitive remarks comes in his later *Durée et Simultanéité*, which in general we have already treated. There, at the beginning of chapter 3, he describes the continuity of our inner life as that of

> a flowing or of a passage, but of a flowing and of a passage which are sufficient in themselves, the flowing not implying a thing which flows and the passage not presupposing any states by which one passes: the *thing* and the *state* are simply snapshots artificially taken of the transition; and this transition, alone experienced naturally, is duration itself. (M, 98)

(Cf. his own English summary of a lecture he gave in 1911 (M, 947 – 8).) 'There are no things, there are only actions,' he insists (CE, 261), and later, in a quite different context, 'the real is mobile, or rather movement itself' (MR, 208). But perhaps the most definite statement of all is one italicised by Bergson himself: 'There are changes, but there are underneath the change no things which change: change has no need of a support. There are movements, but there is no inert or invariable object which moves: movement does not imply a mobile' (CM, 173 (147)).

These passages, the last of which is followed by one of Bergson's main discussions of substance, certainly suggest a direct defiance of Aristotle. Yet there is a certain pervasive ambiguity in his position, perhaps most briefly brought out in the penultimate quotation: 'the real is mobile, or rather movement itself'. Well, which? Or is movement itself mobile? Yes, when he talks of 'a movement of movements' (CM, 175 (148)). Writing in retrospect much later Bergson treats such phrases as an early and inadequate expression of his thought (CM, 84 (72)); but the phrase is relatively late (1911), and he never, I think, repudiated it as anything worse than imprecise. We are told there are 'no things which change', and change needs no 'support', but this is immediately glossed over by the statement that 'there is no *inert or invariable* object which moves' (CM, 173 (147); emphasis mine). Is Bergson saying there is nothing which moves, or simply making the much weaker claim that there is nothing permanent? And how is this claim related to the claim he certainly makes very frequently that the division of the world into objects is an artificial division made by agents for practical purposes? This last claim can be given a fairly innocuous interpretation. We have words like 'hand', 'wrist', 'forearm', 'elbow', and it is plainly because of our practical interests that we don't have a word for, say, the wrist plus the lower half of the forearm. It is equally for us to say where the wrist ends and the hand begins, the sort of point on which scientists and laymen may have different views, and which is connected with the general issue of vagueness. That it is we who carve the world up in this way is hardly controversial, and Bergson is surely saying something more.

Bergson attributes part of the trouble to our reliance on sight, which he thinks is particularly given to interpreting the world in terms of objects which preserve a relatively constant form while moving around. He doesn't ask why sight should behave in this way, beyond saying that it is useful for us that it does so. If we turn to hearing, he thinks, we shall find it easier to perceive movement and change as independent realities, and he brings in the example of the melody, which we have already considered from another point of view, that of the unity of duration, in chapter 2. Here we have 'the clear perception of a movement which is not attached to a mobile, of a change without anything changing' (CM, 174 (147)). The argument which he immediately brings in to support this is at first sight an odd one. He insists, as we saw earlier, on the unity of the melody, which is not just 'a juxtaposition of distinct notes' (ibid.). But what is this supposed to show? It is true that on the dismissed alternative there would not be a 'thing' that changed but a succession of 'things', if anything. But what does that show about whether the melody, granted that it is a unity and not a multiplicity, is a 'thing' that changes or moves?

What the comparison with sight suggests is this. When we see a moving object we see something which has the same form but appears now in one place and now in another. But since it has the same form each time (it is a cricket-ball or whatever) it doesn't seem to matter to it whether it moves or not. Movement is 'taken as super-added to the mobile as an accident' (CM, 173 (147)). Presumably the same would apply to objects which changed in, say, colour. With the melody on the other hand things are quite different. Without the variation in pitch etc. the melody simply would not be itself at all. There is no one constant thing, analogous to the cricket-ball, which undergoes changes, though neither is there a constant succession of different things, since these would no more constitute a melody than would a note heard by me followed by a note heard by you.

This certainly suggests that the change in pitch involved in the melody is not an accident, not something belonging to a subject that could get on perfectly well without it. But not all changes are accidental in this sense anyway. A man goes through a cycle of development from infancy to old age, unless he dies before old age is reached, and he would not be a man, but an angel or something like that, if he did not. It can well be essential to something that it changes. Now a man is not the same sort of thing as a melody. A man exists as a whole at any one moment, even though he is only a man if he has a certain history. The fact that a man must have a certain history if he is to be a man at all no more implies that the history is *part* of him than the fact, if it is one, that I must be the son of my father to be myself implies that my father is part of myself. Having the history is a property of the man and having my father is a property of me, but the man and I both exist completely at the present moment. This is what is not true of the melody – or, at least, so it appears. But perhaps a melody is a sound which changes in pitch, as a man might be, roughly, a body which changes in height? We might answer that a man can be recognised as such at any one moment, which a melody cannot. But this is not conclusive. If the melody has a rich symphonic texture it might well be recognisable on the playing of a single chord from it. Yes, but if that is all that is played the melody never reaches a completion. If a man dies prematurely his life may not reach completion, but he was complete at every moment of it.

So men are different from melodies, and the fact that change can be essential to something is not confined to things like melodies but can apply to things ordinarily accepted as enduring substances. Since both men and melodies therefore differ from cricket-balls in this way, i.e. in essentially involving change, what distinguishes melodies from cricket-balls here cannot be something that distinguishes melodies from substances as such; it cannot show that melodies are pure change without anything there that changes. Change is non-accidental, i.e. non-contingent, in the case of a melody as it is in the case of a man, but this does not imply that in either case it is non-accidental in the sense of not presupposing a substrate. But even if change were found to be non-accidental to melodies in a way that distinguished them from both cricket-balls and men, this would hardly suit Bergson's purpose, since he wants to make a point about all things, melodies, men, and cricket-balls alike; melodies are simply taken as an example.

5. Pure change?

However, someone might argue like this: I have put men and melodies together, despite their differences, because they both involve change essentially, unlike cricket-balls, which don't. But since men are presumably substances, at least for anything we have said so far, the fact that change is involved essentially in something does not show that that thing is 'pure change'. Therefore Bergson's example does not do what he wants it to do. However (the objector continues), the fact that one argument does not establish a certain conclusion does not show that no argument will. Now I have admitted that melodies are different from men because they are not complete at a moment; they are more like events than substances, and perhaps it is just for this reason that they can be called pure change. If so, it is true we have not shown that everything is pure change, but at least we will have shown that some things are, so that the notion is a coherent one and Aristotle's scheme of categories is not all-embracing.

So argues the objector, and at least some of what he says is surely right. It is very hard to make Aristotle's scheme of categories cover as wide an area as one feels it should, and events seem to have no place in it. A few paragraphs ago I toyed with the idea that a melody is a sound which changes in pitch. I dropped it then because it did not alter the fact that a melody is not complete at a moment. Let us now return to it.

Suppose I play 'Home Sweet Home' on an oboe. When does the melody exist? Smith, an adult, exists fully at any time during his life, and so does the adult he is, though he was not always an adult. But the melody never exists fully in that sense. Its existence covers time in a way the existence of a continuant, which is *in* time, does not. If the melody is a sound which varies, does the melody vary? Only, I suggest, in the sense in which France, if it is flat in the west and hilly in the east, varies in hilliness: parts of it are hilly and parts flat. Similarly parts of the melody are high and parts low. The melody is more like Smith's life than like Smith, though really it seems to be a type, of which what we hear are tokens. This is like, but not quite the same as, saying that man is a universal. Smith is a man, but my performance on the oboe is not *a* 'Home Sweet Home', though it is *a* performance of it. The sound is rather harder to classify. Sounds are not as dependent on structure as melodies are, and so the sound does not seem to need temporal parts in the same way. If the sound is indeed something which never exists all at once then the melody can presumably be a sound which varies. But if the sound is rather a sort of continuant, but one accessible only to one sense, then to say the melody *is* a sound which varies would seem to be a shorthand for some more complex expression, bringing out that the sound varies and the melody is the sound but the melody does not (in the same sense) vary. The melody is not strictly identical with the sound, but each temporal part of the melody coincides in a certain sense with some physical part of the sound.

The discussion of the last two paragraphs has given us two candidates for being instances of 'pure change', melodies (and sounds in so far as they are considered as events) and sounds considered as continuants accessible to only one sense. Let us consider these in turn.

If a melody is an event, and still more if it is a type, it does not itself change at all, at least in any relevant sense. Does it, though, somehow contain or consist of 'pure change'? But what would this be? How would one tell one change from another, or one type of change from another? How could a melody have structure if it had no content? Structure of what? We seem to need something like sound, for a melody seems to be a structure of sound (or sounds). If the sound is itself an event then to call the melody a sound which varies is really to give no more than a definition of it. The sound will 'vary' only in the sense in which France varies between west and east, and we should have given no more than a verbal account of the change we are concerned with. So we are led to our second candidate.

The sound as a continuant would be something that varies in the strict sense, not just something different in different parts of itself. But what is this sound? It is tempting to give it a material base in terms of air waves etc. We could then say it was air in motion, and air is solid and substantial enough to satisfy our demand for a subject which changes. But though there no doubt is some air there which is vibrating this is purely contingent. *Qua* sound all we have is something accessible to one sense only. We knew about sounds long before we knew about air waves, so our concept of sound cannot itself make any reference to air waves. Someone might say that though we have not always known about air waves or other relevant waves we have always thought of sound as having *some* physical basis, so that our concept of sound has an existential quantifier built into its analysis, as for Martin and Deutscher (1966) our concept of memory carries a reference to *some* physical mechanism and for Davidson (1967) our concept of causation carries a reference to *some* law. But there seems little reason to think this true in the case of sound at any rate. That a sound should occur which had no detectable physical basis may offend our scientific intuitions, but leaves our logical ones intact. A sound that is only accessible to the sense of hearing is in a way 'insubstantial', though it can still be objective, or at any rate intersubjective; it can be such that anyone suitably placed could hear it. It is also substantial enough to carry properties. It can be high or loud at one moment and low or soft at another, and would change from one to the other in the sense that it existed as a whole at each moment. We need not here, I think, go into questions of individuation or Strawsonian questions about how far a purely auditory world could give us concepts of objectivity or externality.

The case of sound involves qualitative change. But how about locomotive change, or motion proper? Bergson is impressed by the progress of modern physics, and finds confirmation for his own views in it: 'The more it [physical science] progresses the more it resolves matter into actions moving through space, into movements dashing back and forth in a constant vibration so that mobility becomes reality itself' (CM, 175 (148)). This colourful picture of 'movements dashing back and forth' is not perhaps the most precise of descriptions, but the general point is clear enough. Objects are reduced to molecules, molecules to atoms, and atoms to 'electrons or corpuscles' (ibid.), which are strange things by any account, lacking not only solidity, at least in the normal sense, but even normal principles of individuation. He goes on to describe the mobile in the case of sight as a coloured spot, which reduces to a series of vibrations. To say the coloured spot 'reduces to' the vibrations suffers from the same defect as saying the sound reduces to the sound waves, but we can pass that by. What is for most people a stumbling-block of modern physics, that it seems to abolish solidity and material substantiality, is grist to Bergson's mill, which Einstein's '$E = mc^2$' apparently reinforces – though it hardly does: to say that the piece of uranium in my hand could vanish into an explosion if suitably stimulated is not to say that it *is* an explosion, or that my holding it could better, or even as well, be described in terms of explosions. How closely the language of modern physics really resembles a 'process' language I am not qualified to say. But it is a language invented for special purposes, and presumably could not be understood except by someone already knowing a natural language.

6. The substrate and its role

Earlier I asked whether Bergson was really concerned to deny that there was any subject of movement, any mobile, or was making the weaker claim that there was nothing permanent underlying movement. Let us now examine this distinction a bit further. There has been a tradition in philosophy going right back to Parmenides that substance or that which is real is permanent. Anything which perishes must have a basis in that which does not. The standard way for a material object to perish is by dissolution into parts e.g. atoms, which survive the change but are re-arranged in space. We normally think of an object that moves as taking its parts with it, as a cricket-ball does, but some objects seem to move by replacing their parts, which themselves remain almost stationary; a sea-wave is an example. A wave can perish without losing any parts, if it just flattens out, but this would involve some re-arrangement of its parts. Because of the constant flux of its parts a wave is not a very standard example of a material object; it seems more like a mere phenomenon; but it is merely an extreme case of something familiar enough since Locke pointed out that biological entities are constantly changing their matter. The wave is an extreme case because it changes its matter whenever it moves, which most waves do all the time, and it moves *by* changing it. A man, though, as I said above, he must change if he is to be a man at all, need not strictly change his matter, though of course he does. Another feature of waves is that they not only have vague edges (but so do many things, such as hands, or islands in a marsh) but are delimited by their form. In two dimensions, though not in the third, the matter just outside them is qualitatively the same as the matter inside them, i.e. water or whatever. But this should no more stop a wave being an object than it stops a mountain, which is similarly continuous with the surrounding plain. There are real distinctions between objects – a cricket-ball is really distinct from the air around it – but how we divide the world up into objects is a pragmatic matter, at any rate when no void is involved, and often when there is: a galaxy can be an object.

A wave then does have a material substrate even if this is always changing. (I am thinking of ordinary waves, not things like gravity waves, which might need special treatment.) But what is the role of this substrate? If a cricket-ball is made of cork, when the cricket-ball moves the cork moves. But when the wave moves the water does not move, or not in the same direction, but first a piece of water has the property of being raised above its surroundings and then a neighbouring piece of water acquires that property and the first piece no longer has it. We could easily get rid of even the transverse movement of the water by replacing the property of being raised above its surroundings by some other property, such as having a certain colour or temperature. It is clear enough how this can be described by saying that 'real movement is rather the transference of a state than of a thing' (MM, 267), to quote the last of four propositions in which Bergson sums up this part of his philosophy. But we have seen that we can also describe the wave as a thing which moves, even though it can only do so by changing its material *pari passu* with its movement. What then is the advantage of speaking in these terms? One advantage is that it provides a principle of constancy or continuity that the other view does not seem to provide. For, if we talk of the transference of a state, what exactly does this involve? What is being transferred? The state of being humped, in the wave case, i.e. a certain shape? But would we equally talk of a transfer of state whenever a certain body of water ceased to be humped and another body somewhere else became humped? Would we not insist at the very least on these two events being simultaneous, or having some systematic temporal relationship, e.g. such that the time interval was proportional to the space interval between these events? But why should we insist on this, if not because otherwise we wonder what happened to the humpedness in the meantime, or whether we still have the same humpedness? Now what do these questions imply? Humpedness is a quality, a universal. It can hardly be anywhere 'in the meantime' since it is not in time at all, and the 'same' humpedness can mean only something like humpedness of the same degree. The questions are evidently trying to treat humpedness as an Aristotelian accident, something peculiar to one individual substance, e.g. body of water. Aristotle apparently believed in such things. (See his *Categories*, chapter 2.) But they are hard to make sense of, and could not by definition be transferred from one substance to another, nor exist without a substance to inhere in.

No doubt it will be replied that this is to take 'transference of a state' too literally, too much in the way Bergson is trying to escape from. A state is not a sort of thing, and all the constancy and continuity we need is provided if we think of the whole body of water, the sea or whatever, being humped now in one part of itself and now in another, these parts being indeed systematically related, so that we can apply laws of nature to the phenomenon; nothing is being transferred or moves but something, the sea, is changing qualitatively. We can then extend the idea to cover all motion, replacing the sea by the universe as a whole, which is constantly changing in this way in different parts of itself. (Cf. MM, 258-9.)

This reply seems fair enough. But it raises some questions. First we have not got rid of the mobile in favour of pure movement. We have rather got rid of movement itself in favour of qualitative change. We can only keep movement by talking of the transference of states and taking this literally, a state being a kind of thing. Either there really is a transference or there isn't, and, if there is, there is something which is transferred. *Matter and Memory* insists that motion is qualitative and not spatial, but Bergson does not seem to want to say it is not motion at all but something else. Secondly, and more importantly, we have still got qualitative change, which requires a substrate, at least for anything we have said so far. We have not reduced the world to pure events, still less to 'a movement of movements' (CM, 175 (148)), and it was states as well as things that were described as artificial 'snapshots' of the flux (M, 98).

7. Permanence

But there is more to be said. Going back to the strong position that there is nothing that moves and the weaker one that there is nothing permanent, someone might say that the weaker in fact implies the stronger, i.e. that if there must be a substrate at all there must be a permanent one. Take a flame, which is like a wave in constantly renewing its material but without the complication of moving. The flame renews its material by shedding some molecules and acquiring others. But what about these molecules? To know they have been shed or acquired we must be able to identify them. So consider one of them. It must persist at least long enough to be shed or acquired. Perhaps it can itself shed or acquire atoms, but how far can this process continue? Flames are made of molecules, and molecules of atoms, and – must we not stop somewhere? Perhaps atoms could in principle, *pace* modern science, be made of 'bare stuff', as they were for the Greek atomists. Could an atom then replace some of its matter? But how could it do so without a bit breaking off, this being simply a sub-atom? In fact of course atoms are now said to be made of electrons and nucleons, but these are rather strange things, 'waves of probability undulating in nothingness', as Russell somewhere calls them. Evidently the mechanical model starts breaking down at this point. But does this really matter? Suppose we stop with atoms and say that atoms cannot change their material; it is essential to an atom that it has the material it has, though this is only a way of saying it does not make sense for an atom to change its material, since its material can only be identified in terms of the atom itself. Must the atom be permanent? Well, why should it not simply vanish, and another one come to be? Parmenides and many philosophers since would regard this as unintelligible, but it breaks no logical law, and in at least one modern theory (the continuous creation theory of a few decades ago) particles could pop into existence if not out of it. Such a theory is no doubt uncomfortable, but from the point of view of our present concerns one might ask how different it really is from that which lets atoms dissolve into 'funny' things like electrons.

The weaker position then, that there is no permanent substrate, does not imply the stronger position, that when there is movement there is nothing that moves. A further point is this: can one deny that there is a permanent substrate only at the price of holding that matter can pop into or out of existence? I think this price need not be paid. Take two Greek atoms, two bits of solid stuff, and let them come into contact so that they merge into one larger atom. The Greeks would never have let them merge, though it is hard to see how they could avoid merging if they really came into contact, with no void in between. But when they did so form a larger atom the material would have changed from being the material of the two old atoms to being that of the new one, if I was right in saying just above that such material could only be individuated in terms of that whose material it was.

8. Substance

It is time to take stock of where we have reached. Let us allow that there is no permanent substrate, since we are not Greek atomists. There need be no permanent objects; things in the world can merge and separate like drops of water in a fountain, though in practice many of the drops are relatively long-lasting and many more appear to be more long-lasting than they are; that they do so appear is no doubt useful from a practical point of view, as Bergson insists. But, while we can agree with him in seeing matter apparently dissolve away at the hands of science, we need not follow him in talking of the apparent vehicle of movement as 'a coloured spot which . . . amounts, in itself, to a series of extremely rapid vibrations. This alleged movement of a thing is in reality only a movement of movements' (CM, 175 (148)). Leaving aside the question how secondary qualities are related to what underlies them, we can simply object to taking vibrations as ultimate, and to talking of 'a movement of movements'. The question Bergson has not yet answered is this: if there is nothing that vibrates what is the justification for using the word 'vibration' at all? Granted, there may be nothing solid like a violin string vibrating. Perhaps what is happening is this: a light appears briefly at a certain place (let us ignore how we individuate places); as soon as it goes out another light appears equally briefly at a neighbouring place, to be replaced immediately by the first light, and so on. It is a well-known psychological fact that in suitable cases we should see such a phenomenon as a single light oscillating in position. But what we have got here is not 'pure movement', but apparently a light which moves and really two lights which appear and vanish alternately (or a whole set of lights which do so once each, if you like).

What is true is that we have not here anything solid and material. The question how a thing is related to its properties is a large and ancient one, and both Bergson and modern science have shown us that the common-sense notion of solidity is more problematic than a common-sense philosopher like Locke would take it to be. In particular a thing need not be accessible to more than one sense. A ghost that was publicly visible would be as real as a sound. A collective hallucination is such only because there is a wider public that does not perceive it, or because it misleadingly purports to have properties radically beyond those it has. I won't ask whether a thing must be accessible to at least one sense, beyond saying that presumably it must give us some evidence for postulating its existence. But properties, including movement, cannot exist on their own. Even a coloured spot is a coloured *spot*, with a shape, size, and location.

How much of this does Bergson wish to deny? Some of it evidently, but not all of it. Let us return to Bergson and Heraclitus. When he makes this general point about change in the 'Introduction to metaphysics' of 1903 he says there is an external reality which is given to us immediately, and this is mobility. In reissuing this in 1934 he added a footnote saying:

> Let me insist I am thereby in no way setting aside *substance*. On the contrary, I affirm the persistence of existences. And I believe I have facilitated their representation. How was it ever possible to compare this doctrine with the doctrine of Heraclitus? (CM, 305 note 23 (Q, 1420))

To judge by the indexes to *Oeuvres* and *Mélanges* this is, with one irrelevant exception, the only time he ever mentioned Heraclitus, though one might think he had given plenty of reason for the rejected comparison. It is not clear what he takes Heraclitus to have held, but, when he goes on to say in his main text, 'There do not exist *things* made but only things in the making' (CM, 222 (188)), this suggests he is allowing that there are things but insisting that they are perpetually changing. In another footnote added later to an early work, he asks,

> From the fact that a being is action can one conclude that its existence is evanescent? What more does anyone say than I have said, in making it reside in a 'substratum', which has nothing determined about it, since, by hypothesis, its determination, and consequently its essence, is this very action?
>
> (CM, 305 note 19 (Q, 1382))

He goes on to refer to the persistence of the past in the present, which we have not yet discussed, and attributes the trouble to a confusion of 'real duration' and 'spatialised time'.

The point that it is vacuous to posit a substrate whose essence turns out to be that whose substrate it is recalls the point I made above about the material of an atom being only individuatable in terms of the atom itself. But it is not clear that this obviates the fact that in the one case what has to stand on its own feet is an atom and in the other case an action, and that atoms and actions have significantly different places in the Aristotelian categorial scheme.

Bergson does believe in substantiality in some sense because it is essentially involved in duration. As Gouhier puts it in his introduction to *Oeuvres*, 'a becoming without being is no more real than a being without becoming', and he goes on to treat the sort of Heracliteanism which Bergson rejects as that for which 'everything flows' means 'nothing remains' (pp.xxii – xxiii, my translation). But what remains? At CM, 175 – 6 (148 – 9) Bergson says that the *substantiality* of change (his italics) is most clearly seen in the inner life. Trouble arises when we analyse this into 'a series of distinct psychological states, each one invariable' and 'an ego, no less invariable, which would serve as support for them'; but 'the truth is that there is neither a rigid, immovable substratum nor distinct states passing over it like actors on a stage. There is simply the continuous melody of our inner life'. But whose inner life? What is supposed to be the difference between mine and yours? It is all very well to say that the division of the flux of experience into objects is something we engage in for pragmatic purposes, but who are the 'we' that engage in it? Surely we do not owe our distinction from each other to pragmatic purposes, for whose purposes could they be? But we will return to personal identity later.

Aristotle's doctrine of categories has always been subject to a double interpretation. It can be regarded as distinguishing substances as the bearers of items in the other categories. Or it can be regarded as picking out anything at all, be it substance, quality, time, action, change, or what have you, and saying what it is; predications of this last type count as in the first category – they give the essence or substance of what the thing is. I don't think Bergson has any quarrel with the second interpretation. It is the first, that on which I distinguished atoms and actions a page or two back, that he seems to want to reject. To accept the second while rejecting the first is no doubt less puzzling than to reject both. But can we really make sense of a universe whose contents are changes or vibrations or actions or qualities which are not changes etc. *of* anything? Must not a change be *from* something *to* something, e.g. a change of colour from red to green? To say just this, however, is not to give us enough, for the 'of' is not the 'of' we are looking for. If there is nothing which first *is* red and then, being the same thing, *is* green, why have we got a change at all? Why have we not simply got a case of red and then a case of green, without change coming into the picture? Change essentially involves inconsistency – another doctrine Aristotle came near to, if without this time very precisely formulating it. The reason why a sweet white thing which becomes black and bitter changes *from* being sweet *to* being bitter, but not *from* being sweet *to* being black, is that it cannot be both sweet and bitter at once, though it can easily be sweet and black at once. This requirement of inconsistency seems to require in turn an underlying unity, the substrate, which cannot be inconsistently qualified at the same time. Of course none of this commits us to accepting the precise details of Aristotle's scheme, which I have already suggested is inadequate in accounting for things like events.

A Bergsonite might reply to this that I am simply begging the question. It is Aristotelian change that requires inconsistency in this way, and the kind of replacement which I have dismissed is just what Bergson wants to maintain. I think there are two answers to this. First, as we have just seen, Bergson does in his rejection of Heraclitus accept some sort of substantiality; it is actions and movements rather than objects that are the substances, and somehow persons must fit into the scheme. Second, it is not clear anyway how far this anti-Aristotelian line could be pushed. We seem to be invited to envisage an account of the world in this sort of way: 'White sweet here at t_1; black bitter here at t_2; . . . .' But if this is to make any sense at all there must at least be some way of identifying 'here'. There can only be *replacement* if there is a tacit reference to inconsistency – bitter cannot 'replace' white –

and what inconsistency can there be without a subject, even if only a 'region', that cannot be white and black at once? Without this how can we talk of *change*, rather than simply of a set of successive times each garlanded with a set of descriptive terms, these sets being different? But this would in any case be the very thing Bergson wants to avoid, the antithesis of his view of change and duration.

9. Things and processes

So let us abandon this extreme flux, the sort of thing I assume Bergson attributed, though surely wrongly, to Heraclitus, and return to an ontology of actions and movements. I have already criticised this by asking what the movements would be movements of. But if this view is incoherent the incoherence, it will be said, does not only belong to Bergson. Apart from people like Hegel and Whitehead, whom I have mentioned already, there are modern writers who defend something certainly in this area. E.M. Zemach (1979) claims that four different ontologies are all on a level: ordinary objects, which have parts in space but continue in time, processes, which have parts in time but continue in space, four-dimensional space-time worms, which have parts in both space and time and continue in neither (he calls them 'events'), and 'types', like man, gold, the elm, which have parts in neither but continue in both. We are only interested in the first two of these, in particular with the second, which has as examples things like 'this noise' or 'the French Revolution'; 'this noise' of course reminds us of the melody example yet again.

Zemach's strategy is to compare the behaviour of processes with that of the more familiar category of ordinary objects in respect of space and time, showing that processes and objects (or things, as he calls them) are duals of each other in that what one can say of one in terms of space and time one can say of the other in terms of time and space respectively. He adds that, unlike Bergson and others, he doesn't claim that process ontology is *the* correct ontology, but that the two ontologies are on a level and each of them self-sufficient, so that one's language could be framed in terms of either. He claims in fact that this follows from the self-sufficiency of the thing and 'event' ontologies, which he reckons to have already shown – or rather, in the case of the 'event' ontology, assumed, saying that 'most contemporary philosophers' allow the language of 'events' is at least as adequate as any other for categorising reality (Zemach 1979: 66). Presumably Bergson is not included among those who would allow this, since such a symmetrical treatment of space and time would clearly be anathema to him for an adequate description of reality, though he might well allow it for scientific purposes. It is ironical that we might want to say it is a 'process' language that is to be allowed for scientific purposes, if the language of physics is indeed such.

However, it is important to ask just what is being maintained. If two languages are such that anything said in one of them can be re-expressed in the other, it does not follow that the languages are on a level in all respects. One of them may be parasitic on the other, in the sense that one could only learn it if one already knew the other. We can talk about the French Revolution and its various phases and stages, its battles, reforms, development, causes, and effects. But could we have the concept of, say, a battle at all without having concepts of things like warriors and weapons? Even if one could analyse these concepts themselves into concepts of processes, and so in a sense by-pass them, it is not clear that the concept of a battle that we would end up with as a result of doing this would be the concept that we have at the moment, i.e. the fact that at the moment we have process-concepts like that of battle does not seem to imply that those same concepts would survive into a thoroughgoing apparatus of process-concepts; and there would still be the question whether such an apparatus could be independently acquired. But, be that as it may, the distinction I am drawing between translatability and total equivalence can be illustrated by reference to a similar situation, that of Goodman's 'grue' paradox. (See Goodman 1954.) That the grue/bleen pair and the blue/green pair are interdefinable is uncontroversial. That they are epistemically symmetrical is anything but uncontroversial.

So even if everything we say about objects could be translated into terms of actions it would not follow that we could start originally from either end indifferently, and still less of course would it follow that the action/movement language was prior – indeed if it did Zemach at any rate would have to be wrong, since for him the languages are on a level. Furthermore it would not follow that the language which has actions and movements as its subject-terms would also have movements among its predicate-terms, as Bergson implies when he talks of 'a movement of movements' (CM, 175 (148)). One might well expect that such a language would have completely different predicates altogether, not just those that apply to objects in the language of objects. Many of the predicates we apply to movements in our ordinary language we do not apply to objects, like 'took an hour', 'was over quickly', while others have focal ambiguity, like 'fast' as applied to cars and journeys, or, on one view, 'is a mammal' in 'the whale is a mammal' and 'Moby Dick is a mammal' (Wisdom 1931).

There is one more ambiguity to be dealt with. I have so far assumed, in the light of remarks like 'There is no inert or invariable object which moves' (CM, 173 (147)), that Bergson is either saying there is no substrate or saying there is no permanent substrate. But there is another possibility, that he means there is a substrate but it is always changing (as against merely not never changing). How far this is a coherent notion seems to depend on how strictly it is taken. Perhaps something can change continuously in, say, colour, shape, size, and position, and can still be re-identified, at least if not too many of the changes are simultaneously discontinuous. But it cannot change *all* its properties at once. Not only is the idea that it could do so dubiously coherent in the case of second-order properties like that of changing, but some of its generic properties must remain constant if it is to be identified at all, for what sense would it make to talk of something which could not be referred to by any property at all? It seems then that this alternative, that there is a substrate but only an ever-changing one, is only a real alternative if it is not taken too strictly.

10. Conclusion

To sum up the last part of this chapter then: Bergson does not want to deny the logical aspect of Aristotle's doctrine of categories, roughly that predicates need subjects. But he does want to deny the metaphysical aspect, that the subjects in a proper description of reality are substances in the sense of Aristotle's 'primary substance', physical objects (including things like atoms). He is not simply saying, though he would agree, that no such subject is permanent. But neither is he saying that every such subject is perpetually changing in the extreme, allegedly Heraclitean, sense of changing *all* its properties all the time, an idea which is incoherent and anyway would only lead, if anywhere, to the cinematographic view which except for pragmatic scientific purposes he rejects. He seems to want the more moderate flux doctrine that everything is always changing in certain respects – certain definite respects, not just that everything is always changing in at least one respect. The definite respects in question, whatever they may be, are those involved in duration. This flux doctrine, however, he often expresses by treating movement itself as the subject or substance, in a way that makes it look as if he is denying the logical aspect of the Aristotelian doctrine of categories. His way of speaking amounts to a 'process' language, which he treats as not only available as a translation of ordinary 'thing' language, but also as the correct substitute for it. He treats it therefore as not just definitionally equivalent to 'thing' language, but as epistemically equivalent, if not prior, to it. The issue on this last point depends on whether he thinks our ordinary language is somehow parasitic on the correct language. But he seems in fact to treat language itself as part of the general pragmatic enterprise of science and daily life – we only need language for practical purposes – and if he does mean this he need say only that 'process' language is epistemically equivalent, not prior, to 'thing' language. But it would be hard to see how it could be the correct language if it were epistemically posterior, i.e. could only be understood by someone who already understood 'thing' language.

How far this preference for 'process' language fits in with Bergson's own practice is another matter. He talks, as we have seen, in terms of images and objects. Perhaps this is because of what he would regard as the natural deficiencies of language. So 'process' language is posterior after all? Not necessarily. The point is rather that language as such is an imperfect instrument. The point is not that 'process' language cannot be understood without 'thing' language but rather that it cannot really be constructed at all. Nevertheless what it tries to say is true, in Bergson's view. There is undoubtedly a certain tension here, but Bergson would not, after all, be the only philosopher whose views on language sat unhappily with his attempts to express his own philosophy in that language.

V

Problems of mind and body

1. *Introduction*

In the last chapter we saw something of the general metaphysical position Bergson adopted, of his attempt to steer a middle course between realism and idealism, and to repudiate 'ordinary' dualism while preserving a dualism of his own. We saw that the system gave a basic role to what Bergson calls 'images', which provided a monistic element tying his whole philosophy very closely to the mind/body problem. The central position in Bergson's thought of this problem, or nexus of problems, is what gives it an idealist tinge, despite his protestations.

Bergson's own dualism might be described, if rather puzzlingly, as one between perception and memory. That the dualism should be expressed not in terms of consciousness on the one hand and an external world on the other but in terms of two activities of consciousness itself is in keeping with the whole spirit of his philosophy, even though I know of no passage where the dualism is described point blank as one of perception and memory. It is between these terms that he insists time after time that there is a difference of kind, not of degree, though there are complications; perception and memory can in a certain sense be equated with matter and spirit respectively (MM, 325), but matter and spirit themselves are said to differ only in degree (MM, 296).

It is hard to conclude after reading *Matter and Memory* that Bergson has any unambiguous and straightforward attitude to the relations between mind and matter. On the surface and explicitly he is a dualist. It is when we dig deeper into the more technical features of his system that his monism shows through. His attitude to monism of the materialist or idealist kind and 'ordinary' dualism is perhaps a further example of his habit of attacking what was in common to two opposite conceptual systems, well brought out by Mossé-Bastide (1955: 277 – 9). (See also Kolakowski 1985: 6 – 7.) What *was* in common in this case would simply be the false disjunction, as he saw it, between seeing no difference at all between mind and matter and seeing them as entirely cut off from each other so that any attempt to bridge the gap between them must fail.

Both perception and memory have 'pure' forms as well as their ordinary forms, but in conscious experience they always occur together, being mixed together in various degrees, despite the fact that in themselves they differ in kind and not in degree. This does not mean, however, that they are mirror-images of each other; they do show a certain symmetry, but only within definite limits, and they must be treated separately.

I have already introduced the notion of images and said something of the role they play. This comes out rather strikingly when Bergson discusses and defends the notion of unconscious representations, which might seem an obscure notion if not a contradiction in terms. For, it might be said, what could a representation be except a representation *of* something *to* a consciousness? A map is a representation, and is not itself a conscious phenomenon, but this is only because we think of it as made by, or at the very least being usable by, a conscious being. Perhaps one could argue that a DNA molecule is a representation. To discuss that would take us astray, but it is unnecessary here because the context plainly concerns something different. Everyone admits, Bergson says, 'that the images actually present to our perception are not the whole of matter. But, on the other hand, what can be a non-perceived material object, an image not imagined, unless it is a kind of unconscious mental state?' (MM, 183). Unperceived objects, he goes on to say, are 'so many perceptions absent from your consciousness and yet given outside of it' (ibid.). It is typical of Bergson's cheerful assumption that he represents common sense against the distortions of philosophy that he introduces this bit of the discussion by saying, 'we may even say . . . that there is no conception more familiar to common sense' than that of an unconscious representation (ibid.).

Here again only a very unsympathetic critic would accuse Bergson of confusing two senses of 'image', the ordinary one and his own technical one, which he would regard, I think, as a development of the ordinary one. All we know of the world is through our perceptions of it. But if the words 'of it' suggest that the object has some hidden reality of its own independent of our perceiving it then how could we know of this? How can any object that we are to know of be more than a 'permanent possibility of sensation'? Yet, as we saw in chapter 4, we must not embrace idealism with its suggestion, by contrast, of something that might be there but isn't. One might think that phenomenalism would better represent Bergson's position. But that too would make the success of science an accident. To say that images can exist unperceived is not a mere *façon de parler.* Unperceived images are real enough, but they do not have unperceivable aspects. What Bergson does not allow for is that they might have aspects which we could not directly perceive but could know about because they were presupposed by the possibility of our perceiving what we did perceive.

There is a certain tension in Bergson's thought which arises from his 'neutral monism'. The same images belong at the same time to two different systems, and he himself regards this as generating the problem at issue between realism and idealism (MM, 12 – 13). In one system they form my perception of the universe, and centre round one privileged image, my body, slight variations in which can cause large variations in the rest. In the other system they form the universe, influencing each other causally but without any special centre. This is perhaps the most explicit statement of neutral monism in Bergson's work, and it is again typical of his assumption of more support than we might be willing to grant him that he can treat this double role of images as something 'no philosophical doctrine denies' (MM, 13). In the 1911 'Introduction' he talks of treating mental states and brain states as 'two versions, in two different languages, of one and the same original' (MM, xv), though there he talks of it only as one view among others which is held, without claiming it as his own.

2. *Perception*

The tension arises concerning perception. Every perception involves duration and therefore memory (MM, 325), but 'pure' perception is perception free from any admixture of memory (MM, 39). Pure perception is an ideal limit (MM, 325), something which exists only in theory rather than in fact (MM, 26); since it involves no memory it involves no duration and is instantaneous (MM, 325). It is 'the lowest degree of mind' (MM, 297) and 'is really part of matter, as we understand matter' (ibid.). In fact perception differs from matter only in degree, and pure perception stands to matter as part to whole (MM, 78). But Bergson is not an idealist; matter is not unreal. How then can pure perception be a part of matter and yet be something that exists only in theory and not in fact? And, if perception is something differing from memory in kind and not degree, how when separated from memory can it become something unreal, which exists only in theory? Is it not strange that 'pure' perception turns out to be not really perception at all? How in fact is perception related to what it is perception of?

In the last chapter I suggested how an image might become a centre of spontaneity, perceiving other images by reflecting them differentially. Let us now develop this a bit further. The idea that for an image being and being perceived differ in degree suggests an inert image, or object, which by some process gradually becomes perceived, but what we have just seen about pure perception being part of matter suggests again a difference of degree but this time between being an object and being (not perceived but) perceiving, either an act of perceiving or a perceiver. This is what makes problematic the relation of perception to what it is perception of. One thing Bergson insists on is that perception takes place in the object, in the sense that it is there and not in the brain that the image of what is perceived is formed and perceived (MM, 35 – 8); we do not form an image in the brain and then somehow project it outside. The image of the object is 'of' it in the sense of consisting of it or (partly) constituting it – 'partly', because the image will embody only one point of view. Even if the object is the image they need not be strictly identical, any more than I am with the baby I was – which incidentally answers the objection that objects are not permanent whereas images are, if memory is to survive in the unconscious (MacWilliam 1928: 284). This image is *chosen* from indefinitely many others to become part of my perception (MM, 36). But surely, one might think, it either is chosen or it isn't; how could that be a matter of degree? The question of degree comes in rather when we consider the distinction between pure and ordinary perception. Ordinary perception involves duration and memory, in ways we shall have to consider later, but it can involve them in varying degrees, and the duration involved can be indefinitely short. At the limit it vanishes altogether and we have pure perception – which, however, is only pure in the sense of being unmixed with memory, but ceases to be a conscious experience. (Cf. MM, 80.) That is the sense in which it ceases to be perception, and ceases to exist as perception in anything but theory. But it does not cease to exist altogether, for the image which it is is still there but is now simply matter. What is not so easy to see is how this image which is the pure perception relates to the image which is the object perceived. From what we said about images in chapter 4 we might expect the former of these images to be the body or part of it; for, however much perception takes place in the object, in the sense that the object itself has a direct part to play and does not act merely through a surrogate mental image in the mind, presumably the body must come in somehow and its location must be represented in the location of the perceiving, i.e. the answer to the question where the perceiving occurs must make some reference to where the body is. What Bergson actually tells us is that the object, the light-rays from it, the retina, and the nerves etc. 'form a single whole' (MM, 37 – 8). This somewhat opaque remark can perhaps give us a hint that we are ignoring the result we came to in the last chapter, that Bergson is after all a process philosopher; I have been talking in terms of objects and the body, but should we rather talk simply of the process of perceiving?

As a way out this would face whatever objections face the process philosophy, but let us see more of what it amounts to, keeping, however, to the language of objects for the moment for simplicity of exposition, since it is from that language that we are starting – or so I have claimed. One thing the long discussion of perception at pp.26 – 69 of *Matter and Memory* makes clear is that, as we saw above, Bergson is rejecting a view whereby we start with sensations, which are given to us by our senses but which we regard as unextended and unlocated and then somehow use to tell us about external objects, by some process which is inexplicable. This view is in essentials the same as the traditional empiricist view where we start from sense-data and somehow use these to tell us about objects. Actually, Bergson insists, sensations are located in parts of the body, but because they are only rather vaguely located we, or rather psychologists who know about the functions of the brain and nerves, first try to re-locate them in the nerves or brain and then, since this leads to problems about how they get projected out again, push them out of the brain again and treat them as unextended. They then face the impossible task of showing how they take on extensity as representations of an extended outer world. (MM, 62 – 3. Cf. MM, 311 – 12.)

In making this general criticism Bergson is surely right. 'My belief in an external world does not come, cannot come, from the fact that I project outside myself sensations that are unextended: how could these sensations ever acquire extension, and whence should I get the notion of exteriority?' (MM, 42). But what does he put in place of the rejected view? Briefly, that sensations are located in the body and perceptions are located in the object, so that they are quite different from each other. (Cf. MM, 63.) It is the second half of this that primarily concerns us, but what does it mean? The most obvious interpretation might be that we perceive objects themselves, not surrogates for them, in other words a direct realist view. This is further suggested when he accuses the psychologists (and philosophers) he is attacking of thinking that perception could only be in the thing perceived if that thing itself had perception (MM, 62). Evidently he thinks himself that the thing itself does not have perception, i.e. does not perceive, and indeed it would be odd to think otherwise (ignoring the irrelevance that what is perceived might happen to be itself a living creature). But the trouble is that, as we have seen, he rejects realism along with idealism. Objects are images, and, though they can exist unperceived, they cannot transcend perceivability in any respect. Also the same image belongs in two different systems, in accordance with his 'neutral monism'. This connexion with perception does not involve the object in perceiving, but we have also seen that matter is the limit of the series of perceptions arranged in order of increasing 'pureness'. This suggests that the image which is the object and is also the 'pure perception' would, if it were to be perceived, become itself a perception, i.e. an experience of perceiving, which raises the question we started off from: how is perception related to what it is of?

This is what made me suggest that perhaps we should talk simply in terms of the process or activity of perceiving, treating talk in terms of objects perceived as somehow secondary. The result would presumably be something like this: an act of perceiving occurs involving an image. This image belongs in a system where it is related by causal laws to other images, and also belongs in another system where it is related by what might be called (though Bergson does not use such a phrase, I think) relations of co-consciousness. Its appearance in this second system, suitably garnished with features provided by memory which we have not yet come to, just is the act of perceiving, as its appearance in the first system is the being of the object. If this is indeed Bergson's view it faces objections, notably concerning the relation of co-consciousness, which suggests that there is nothing which *does* the perceiving and that the self is simply a bundle of experiences, or perhaps of actings – but can we have actings without actors? As for the other system and its treatment of objects, it faces the objection about the wind I mentioned when first introducing images in chapter 4.

In the last few paragraphs I have presented a partial interpretation of Bergson's view of perception based on process philosophy and neutral monism. But this interpretation does not spring unambiguously from the text. It creates difficulties for the unity of the mind, but difficulties will exist for Bergson anyway, and indeed for all of us. More importantly it ignores the undoubtedly dualistic elements in his thinking, or anyway downgrades them, for he clearly contrasts mind as the subject of mental attitudes with matter as their object in the case of perception. Also pure perception is something theoretical and unreal; it is 'a vision of matter both immediate and instantaneous' (MM, 26), which does not occur. (Cf. MM, 170.) But the matter which pure perception collapses into is real enough, since we have seen that images can exist unperceived. What we seem to need is that pure perceptions can occur but when they do are no longer perceptions, just as a baby can grow old but when it does is no longer a baby. The contrast between perceiver and perceived will be given by the contrast between the two groups of images which the given image will belong to in the two systems.

3. Memory

But perception involves duration and memory, so now let us turn to memory. Bergson starts his second chapter with one of his most famous distinctions, between two kinds of memory, which he introduces under the names 'motor mechanisms' and 'independent recollections' (MM, 87), or, as we might call them, habit memory and picture memory. Despite being followed here by his arch-enemy Russell his account of the distinction has been described by R.F. Holland, as 'so patently full of misconceptions and so ingenuously pressed into the service of a preconceived metaphysical dualism that one stands amazed at Russell's confidence in it' (Holland 1954: 480). Holland does not elaborate this harsh verdict, but it is true enough that what Bergson says about memory plays a major part in his general metaphysical system, and it is fair to add that he was not primarily concerned with the sort of conceptual analyses that occupy modern analytical philosophy. At a certain level the distinction is fair enough. There is plainly a difference between remembering how to ride a bicycle and remembering one's sixth birthday. The trouble comes in what we try to do with the distinction, especially if we suppose that the occurrence of mental pictures gives us much help in analysing our knowledge of the past. Bergson is inclined to discount habit memory because it does not represent the past but acts it, so that 'if it still deserves the name of memory' this is 'not because it conserves bygone images, but because it prolongs their useful effect into the present moment' (MM, 93). (Cf. MM, 195, where 'habit' is contrasted with 'true memory'.) Evidently anything worth calling memory is something which represents the past by an image (but we shall have to complicate this later). When a dog wags its tail to welcome its master this counts as habit memory because it presumably does not have an image of its master which it invokes to compare with its present perception (MM, 93). Presumably the point is that the dog has acquired the habit of wagging its tail when its master appears, though Bergson goes on to say, rather oddly, that the recognition consists in 'the animal's consciousness of a special attitude adopted by his body' (ibid.). A few pages earlier (MM, 89) Bergson takes the example of remembering a lesson one has learnt, as against remembering some given reading of it. But the examples are not parallel. The dog is not remembering how to wag its tail, or at least this is not the point of the example, nor even remembering *to* wag it, but is remembering its master. Bergson seems to treat any remembering that does not involve a conscious mental picture as habit memory.

One of the things Bergson most wants to attack is the idea that memories are stored in the brain, an idea he thinks fatal to his dualism, since memories must be independent of the body. The deep study that he made of the literature on aphasia is largely directed to this end, though we might now think that his target, the view that individual memories are stored in a one-to-one fashion in the brain, is rather a straw man. He attributes this view to a confusion of his two kinds of memory based on the study of mixed or intermediate cases, for philosophers, he alleges, like to treat phenomena as simple and so treat the cerebral mechanism which is responsible for the motor habit as also 'the substratum of the conscious image' (MM, 104). This leads him to a discussion of recognition, where he uses the facts about psychic blindness to show that recognition cannot consist simply in the calling up of images to compare with the present perception (MM, 105 – 18). Psychic blindness, he tells us, takes two forms. In one case the patient can form memory images of objects but cannot recognise the originals when confronted with them. Here what has been lost is a motor habit associated with the original perception; when we perceive the thing again we cannot react in the relevant way (e.g. find our way around in a house we know). In the other case all ability to form memory images is lost, but there is still partial recognition because the patient can still recognise objects as belonging to certain types; he fails to recognise his wife, say, but does recognise that it is a woman he is confronted with. Bergson concludes that there are two types of recognition, one of which consists in the preservation of a link between perceptions and certain motor habits, and the sense of familiarity is the consciousness of this link. Perhaps this is what he meant by the dog's consciousness of its bodily attitude (MM, 93). The second type we will return to in a minute, but first a word about the wife/woman case (MM, 109). Normally I would recognise my wife because I had seen her before, but I could recognise her as a woman without having seen her before, or even having seen any woman before, provided I had an adequate concept of woman. In principle I could do the same with my wife, if I had had a sufficiently accurate description of her. Here we have a distinction among kinds of recognition that Bergson does not seem to make, at any rate in this context. Neither kind need involve images, but the distinction is important because the second kind amounts to the application of concepts, which plays a major part in perception, which, as we saw, he thinks involves memory.

The second type of recognition Bergson does distinguish is more full-blooded, involving attention and the regular union of memory images with present perception (MM, 119). He devotes most of the rest of his second chapter to it, and it is here that he brings in the evidence from aphasia and elsewhere to support him. 'Is it', he asks, 'the perception which determines mechanically the appearance of the memories, or is it the memories which spontaneously go to meet the perception?' (MM, 119). His own view is the second. His general strategy in using the evidence of aphasia is simply to argue that the symptoms do not correspond to their known causes in the way they would have to if memories were stored in the brain. The symptoms take certain definite and systematic forms. Certain classes of words are lost before other classes, e.g. proper names before verbs; but the causes are random lesions or strokes, and this randomness is not, or not adequately, reflected in the symptoms, which rather take the form of functional disorders. In the related phenomenon of amnesia memories which disappear after brain damage are often later recovered. I cannot claim much acquaintance with modern work on aphasia, but so far as I know nothing has appeared that would worry Bergson. A recent survey emphasises that units like words or sounds are not lost once and for all, but that the trouble largely concerns time coordination, i.e. producing the right unit or element in a process at the right time (Lenneberg 1967: chapter 5).

One need not go as far as pathological amnesia or aphasia to find cases where we lose a memory and then recover it, or cannot find the right word for something and then find it. Presumably the pathological cases seem more significant because in the view to be attacked the memory vanishes because it has been destroyed by brain damage, and in that case how could it be recovered? Even if the brain damage were repaired this would not explain the re-appearance of the memory, just as if an object in a vase is destroyed when the vase is broken we would not expect repairing the vase to restore the object. In the non-pathological cases the reappearance need not be so surprising since there was no reason to think the memory had been destroyed in the first place; any sane view must allow that memories can be dormant and vary in how easily they can be recovered, though we have not yet come to Bergson's treatment of this dormancy.

While Bergson's arguments may be effective against the naive object-in-the-vase view of memory-storage he seemed to have in mind, it is not clear how they would refute a more sophisticated trace theory which allowed memories to be stored in different ways, one of which could come into action when another was destroyed, and which stored different types of memory (e.g. words of different grammatical category) in different types of mechanism, so that it would not be surprising if disappearances occurred in a systematic way despite the random nature of brain damage. He makes the valid point, against a certain kind of associationist, that our auditory image of a word cannot be something implanted in the brain each time we hear the word, because then it would be obscure how the image got disentangled from the myriad tones of voice in which it was uttered or other words which were juxtaposed to it in sentences, especially if we do not allow memory to be active enough to engage in an abstractionist search for similarities (MM, 147 – 8). But though this may hold against the associationist view in question it hardly shows that our knowledge of a word is something that could exist independently of the brain.

But does he want to show this? Later (1901; see M, 463ff.) he explicitly dismissed cerebral adaptability as too complex to discuss (M, 483), but allowed cerebral localisability a role, albeit a minor one which he does not specify (M, 483-4). Mental states require brain states: 'it is the human brain that has rendered possible human thought' (M, 487). But they are under-determined by them; the relation is many – one. What the brain does is make possible the sort of mental life required for action; it links the mind to the world. How even this limited dependence on the brain lets the mind be prior to it and survive it is still unclear.

So far we have been concerned with one distinction Bergson makes, between what I called 'habit memory' and 'picture memory'. But is 'picture memory' a fair term? In his 1912 lecture on 'The soul and the body', reprinted in *Mind-Energy*, he again argues against the 'mechanical registration' of memories and contrasts 'the' visual recollection or image of something with the indefinitely many and varied impressions one may have had of it, adding that 'it is unquestionable that my consciousness presents to me a unique image, or, what amounts to the same, a practically invariable recollection' of the thing (ME, 51). This suggests that 'the' image is a type, not a token. The tokens do seem to be pictures, but what form does the existence of the type take? This leads us to another distinction he makes, which I referred to earlier in saying that both perception and memory have 'pure' forms as well as ordinary forms. The start of chapter 3 refers to three processes as having been already distinguished, pure memory, memory image, and perception, which might suggest that pure memory is the same as habit memory, since that is what has been explicitly contrasted with memory image (or 'picture memory'). But this cannot be so. Habit memory belongs to the present (MM, 91, 93), concerns the body (MM, 92), can belong to animals (MM, 93), and hardly deserves the name (MM, 93) – though we must remember that pure perception was not really perception except as an ideal limit. Pure memory belongs to the past and is consequently unextended (MM, 181), interests no part of the body (MM, 179), and is contrasted with habit memory, at least implicitly, when the 'bodily memory' is contrasted with 'the true memory of the past'; even though 'they are not two separate things', they are 'two functions', and 'that a recollection should reappear in consciousness, it is necessary that it should descend from the heights of pure memory down to the precise point where *action* is taking place' (MM, 197). On the other hand each of 'the two extreme forms of memory' he has distinguished earlier is called 'pure' (MM, 103). Evidently the pure/impure distinction cuts across the other one and refers to the admixture or not of perception. But Bergson shows little interest in pure habit memory and 'pure memory' usually refers to the pure form of the other kind.

What then is this pure memory or pure recollection? Bergson raises this question most explicitly in his final summary (MM, 317). As so often, he makes some acute criticisms of the view he opposes before going on to develop his own view. There is a temptation, he thinks, to regard perception and memory as different only in degree because a memory is simply a weak revival of a perception which is itself simply a sensation. This sensationalist view, on which associationism is grounded, is one of his commonest targets, as we have already seen in the case of perception. But even on the psychological level, apart from its metaphysical inadequacy, this won't do, he thinks, because then there would be nothing to stop us taking 'the perception of a slight sound for the recollection of a loud noise' (MM, 318 – 19), which we never do. Furthermore we could not refer such a weak sensation to the past unless we already had 'the representation of a past previously lived' (MM, 319), nor could we rediscover its date nor explain 'by what right it reappears at one moment rather than at another' (ibid.). The force of the last point may be less obvious, but the general criticism is clear enough.

> The truth is that memory does not consist in a regression from the present to the past, but, on the contrary, in a progress from the past to the present. It is in the past that we place ourselves at a stroke. (MM, 319)

The nature of the progress is again not obvious, but what is obvious is that we must start with a concept of the past. The 'progress' in fact involves the difference we are seeking between pure memory and ordinary memory. Pure memory is not conscious; recollection only becomes conscious when it 'descend[s] from the heights' (MM, 197). Memory retains 'all our states in the order in which they occur' and in that sense is '[c]o-extensive with consciousness' (MM, 195), but only in that sense. What is remembered is conscious experiences, but the memory of them is not itself conscious until 'from the sensori-motor elements of present action . . . [it] borrows the warmth which gives it life' (MM, 197). All this suggests Bergson is well aware of the problem that a mere picture cannot account for pastness, whether or not he can solve it.

In a minute we shall see a further ambiguity in Bergson's thought, but first a problem: pure memory, like pure perception, is not conscious, and indeed we would normally say that at any one time we have lots of memories of which we are not conscious. Since Bergson associates memory, at least 'true' memory, so closely with images the question what unconscious memory consists in should be a problem for him. It is true that images can exist unperceived, but what hold on reality can unconscious memory images have, since current material objects can hardly provide it? Could he bring in the dispositional/occurrent distinction and say that to have an unconscious memory is to have a disposition to form a memory image? Sometimes he talks of 'virtual recollection' in a way that suggests this (MM, 171); but this won't do entirely, for pure memories are still 'objects' for him, even if only 'virtual objects' (MM, 167). He speaks of memories as like mirror-images of perceptions (most clearly perhaps at ME,

134 – 5). But this is a metaphor he never cashes; he uses the negative features of mirror-images, their intangibility etc., but says nothing of what in the metaphor corresponds to their positive features. Similarly he says memories arise together with, not after, perception of the object in question (ME, 128), and uses this to explain *déjà vu*, but leaves us in the dark as to *what* so arises. Also if experience is holistic and cannot be chopped up, and cannot be repeated because duration is cumulative, it is again hard to see just what is preserved in memory. D. Locke thinks Bergson (along with Russell and Ayer) may confuse the occurrent/dispositional distinction with that between memory of items or events and memory of facts or skills etc. (Locke 1971: 46 – 7). Certainly he ties 'true' memory much too closely to images, which suggests that everything else, dispositional memories and factual memories alike, go together into the same ragbag. He contrasts both 'pure' memory and 'habit' memory with ordinary memory which alone is conscious. So far as pure memory *is* dispositional this may suggest he does conflate pure memory and habit memory, though we saw earlier that he also distinguishes them. But to talk of 'pure, or true, memory' (Locke 1971: 44) as the constrast-term to habit memory is to treat 'pure' as ambiguous between 'unmixed' (with perception) and 'real'. It is hard to know what Bergson would have done with Locke's example of suddenly remembering that today is Sunday while going to work, but perhaps he would have called it a mixture of sundry images being realised together with the implementation of a motor activity (returning home). (On this area see also Ginnane 1960.) In fact two things need saying on Bergson's behalf. First, he is not doing 'conceptual analysis' but singling out two psychological phenomena that might be called memory. Second, he is not committed, as Locke seems to assume, to classifying every example unambiguously under one heading or the other.

The ambiguity I mentioned at the start of the last paragraph is this, that having apparently made clear that pure memory is not a conscious phenomenon he goes on, in his 'Summary and conclusion', to talk of 'planes of consciousness' (MM, 319 – 23), which range between the plane of action and the plane of pure memory. One might try to resolve the ambiguity by taking the planes of consciousness to differ in degree, that of pure memory being at the limit where consciousness vanishes and that of action being at the other end where it is most intense. There may be something in this, though we can only tell after seeing more of the role of action, and Bergson himself expresses the relations between the planes only in terms of wideness, those nearest the pure memory end being the widest. But complete consistency may anyway seem threatened since the extreme which I have referred to as the plane of pure memory he also calls the plane of dream, and dreams, however they are related to pure memory, are certainly in some sense conscious phenomena. But of this more anon.

4. Perception and memory

Now that we have said something about perception and about memory we should turn to the relations between them, for it is only when they are mixed together and are not in their pure forms that we have, at least on Bergson's dominant view, a fully conscious experience.

Consciousness essentially involves duration, and the trouble with pure perception was that it lacked all duration and existed, or would exist if it occurred at all, entirely in the present. This raises again Augustine's question about the squeezing out of the present. Since the present lasts for no time at all, how can it be real? The problem may seem more acute for those with an event ontology, for objects, though they last through time, do not have parts in time but are complete at any one time, as we saw in chapter 4, whereas events have parts in time. (I am not thinking of events in the sense of beginnings and endings, but in the sense of Zemach's processes.) An object must indeed last through time if it is to be real, but it seems possible at least in thought to take a literally momentary snapshot of it, which will, as I have just said, be in a sense complete. This of course is how science considers things in Bergson's 'cinematographic' procedure. But with experiences this seems scarcely possible even in thought. We need not go as far as Wittgenstein's example about feeling deep grief for a fraction of a second, for such a feeling would presumably need a temporal structure; one would feel first one thing and then another, roughly. (See Wittgenstein 1953: II i.) But we could not hear a single sound with a definite pitch, let alone a melody, without hearing it over some period. (See MacKay 1958.) Bergson takes the point even further. An experience of colour is one we can only have because billions of vibrations are occurring in the object, so that 'every perception is already memory. *Practically we perceive only the past*' (MM, 194; his emphasis). He is considering only experience, but if we remember that objects are images, which have secondary qualities, he might argue that even objects cannot be complete at a moment. He might use this (though I don't think he does explicitly) as an argument for an ontology of events rather than objects, though it would depend on his view of objects as having their secondary qualities intrinsically and also on regarding secondary qualities as identical with, or at any rate sharing the temporal nature of, the corresponding underlying scientific properties. For we must distinguish two things: hearing a sound, like seeing a colour or a table, involves time. Because of their physical basis the existence of sounds also involves time, in the sense that they cannot be complete at a moment. But this second point is contingent, as I argued in chapter 4 §5.

Bergson powerfully argues for the need for duration if there is to be experience and brings out Augustine's point (MM, 176 – 7, 193), though how satisfactorily he deals with it is another matter. What he says is that 'the real, concrete, live present – that of which I speak when I speak of my present perception – that present necessarily occupies a duration' (MM, 176). This sounds rather as if he is riding roughshod over Augustine, imposing on the 'ideal present' which separates past from future another present which somehow manages to endure. This would lead to absurd results: is this present, or part of it, both past and present? Does it exist all at once? But perhaps what he means is rather that what I perceive (present tense) is, or includes, the immediate past, and perhaps the immediate future. One might think of comparing the way we are supposed to see the past because of the finite velocity of light – but we are not supposed to see both the past and the present because of that; the theory displaces the moment that is seen rather than showing that something longer than a moment is seen.

There are various ways in which perception involves the past. First, it takes time. As Aristotle said, to see something is to have seen it; there is no first moment of seeing (*Metaphysics,* 1048b23). And as Bergson says, there is no sharp line between memory and perception (e.g. MM, 171). Second, how one sees something will be coloured by what one has just been seeing, just as how one sees a colour will depend on the colours that surround it. Third, a similar influence will be exercised by one's earlier experiences; seeing something for the second time often differs from seeing it for the first. Fourth, often, and it might be argued that in reality always, one must see something for a certain time before realising what it is. Fifth, one can neither see X as Y nor see what X is unless one already has the relevant concepts; this is presumably true even of animals in so far as they really see as against simply reacting to patterns of light. Sixth, in so far as perception is an intentional process, as it is when one sees that something is the case, it involves thought, and we come up against a difficulty about relating thought to time. (See Mouton 1969.) When a thought occurs to one, is this a process that takes time, and if so, how does its temporal structure relate to the semantic structure of the thought, if it has one? Does one, for example, first think the subject of the thought and then the predicate, as one must utter the words of a sentence in some order?

It is not clear how far Bergson distinguishes all these features. He seems to have something like the sixth point in mind, but also the first, when in the 1912 lecture already cited he says that in uttering a long word 'my consciousness presents the word all at once, otherwise it would not be a whole word, and would not convey a single meaning', and goes on to conclude that 'you cannot draw a line between the past and the present, nor consequently between memory and consciousness' (ME, 55). I will return to this passage later, when discussing the reality of the past, but first let us look further at the relations between memory and perception.

It is 'memory above all that lends to perception its subjective character' (MM, 80). (The sentence begins with 'If' but clearly gives Bergson's view.) This evidently means that only when combined with memory does perception become a conscious experience. But what is memory's contribution, and how does it relate to the six features I have just sketched?

The discussions of amnesia and aphasia, directed as they are at Bergson's favourite target, the idea that memories are stored in the brain, suggest a concern with relatively long-term memories and so with the third and fifth features; and though a newborn baby is presumably conscious its experiences must be very different from the perceptions of an adult. On the other hand the emphasis on duration and the constant evanescence of the present, seen in remarks like 'every perception is already memory' (MM, 194), suggest rather the other features.

Along with the idea that the brain is a memory store Bergson constantly attacks the idea that our mental life is somehow compounded from atomistic sensations or atoms of any similar kind. One place we have seen this is when he contrasts our single memory of a person or thing with the unnumbered impressions which gave rise to it. (See ME, 51.) Though he would have done better, I think, not to have insisted on calling the memory a 'unique image' he plainly does not want us to think of our past experiences as surviving as a set of discrete items like photos in an album. One way this helps him in is by allowing the experience which results from the mixture of memory and perception to be one experience, a unity formed by its own duration. This would be a hopeless enterprise if the present experience had somehow to embody a whole host of atomic memories. This presumably is why the memory image which is a current experience is contrasted with non-conscious pure memory, whatever that consists in – in particular it is not a nascent sensation (MM, 174).

5. *Rhythms of duration*

An obscure doctrine, which we have touched on earlier, is that duration has different rhythms for different beings and consciousness has different degrees of tension, the greater the degree of tension the higher being the mental life. This is discussed particularly at MM, 270 – 80. Perception 'contracts into a single moment of my duration that which, taken in itself, spreads over an incalculable number of moments' (MM, 276), though elsewhere it is memory that effects the contractions (MM, 25, 77). Evidently this is intimately connected with the interpenetration of the present and the past, and with the idea that '[q]uestions relating to subject and object, to their distinction and their union, should be put in terms of time rather than of space' (MM, 77).

The general idea seems to be this. We see the successive positions of a shooting star as a single line of fire (TF, 195); in fact if it shoots fast enough we would not even know which direction it was moving in, as with streaks in a Wilson cloud chamber. Similarly, if less obviously, in hearing sounds or seeing colours we contract vast numbers of vibrations into a single experience. An accelerating set of taps will gradually take on the quality of a pitched note. Suppose now that the minimum period of time detectable by our consciousness is 0.002" and that we observe red light for that period. We shall have observed some 8×10^{11} vibrations but we shall not of course observe them individually but shall contract them into an atomic red flash; if we continue to observe we shall have a steady experience of seeing red. (I derive the figures from MM, 272.) Now it seems a contingent matter how much contracting a given consciousness does or can do, and the same conscious being may contract what it is presented with into experiences at greater or lesser degrees of intensity at different times. Five hundred experiences a second represents the most rapid rate at which we can experience the world, the nearest we can come to diluting quality into quantity, and the lowest degree of tension of which we are capable. Given 8×10^{11} consecutive light vibrations we cannot do otherwise than condense them into a single experience, at least in practice, though in principle sufficient effort would enable us to relax the tension so that we experienced the vibrations in ever smaller and more rapidly succeeding packets. (Cf. CM, 220 – 1 (187 – 8) – which seems inconsistent with MM, 273 – 4, where, rather gratuitously one would think, the imagination can divide space but not duration indefinitely.) Towards the other extreme some creature might be able to condense the whole period of an astronomical body so as to grasp it in a single perception, as we do the path of the shooting star (TF, 195). The ideal limit in the first direction would be 'the pure homogeneous, the pure *repetition* by which we shall define materiality' (CM, 221 (187)), the 'imaginary homogeneous time' which *Time and Free Will* claimed was an idol of language (Cf. MM, 274), or in other words pure perception, which as we have seen collapses into matter. The limit in the second direction would be eternity, where the whole of time presumably would be experienced at once. (Cf. CM, 221 (187 – 8), and on all this see the helpful remarks of Čapek 1969, and also Čapek 1971: part III chapters 1 – 6, and Dennett 1984: 67 note 27.)

This raises various problems. As we saw in chapter 2 there is duration in the objective world for *Matter and Memory* and later works. If duration must have some rhythm or other, and different rhythms involve degrees of tension and contraction, and contraction involves consciousness, it is hard to see how the external world can have duration. But as we have seen matter differs from perception in degree, and it may be that Bergson thinks of a totally non-conscious world as an ideal limit which would indeed consist of purely independent (and, he would add, homogeneous) moments which involved no 'groupings' or contractions and therefore no causal interaction. Such a view of the actual world as itself to some degree animate, and not at this ideal limit, would fit in well enough with what he is going to say later about the *élan vital*. The rhythm of the world will then be the single objective duration.

A more pressing problem concerns how we are to compare rhythms in terms of tension or rapidity without some yardstick to measure them by. Perhaps we could at least measure them relative to each other and so arrange them in an ordinal series, or else take the shortest actual intervals that physics finds in the external world (if it does) and use the rhythm based on them as at any rate a practical yardstick. (Cf. Čapek 1969: 302 – 3.) There would be a difficulty in trying to use the hypothetical homogeneous time, for just because it was at the ideal limit any bit of it would contain infinitely many 'vibrations', and how would it be possible to construct units out of them?

Bergson tries to provide empirical support for this idea of different rhythms of duration by appealing to the way in which during a short period of sleep, as measured by our waking rhythms, we have a dream which lasts a long time, as measured by its own rhythms (MM, 275). The topic of dreams is obviously relevant to a theory which attributes consciousness to a blend of memory and perception, and Bergson devoted a lecture to the topic in 1901, reprinted in *Mind-Energy.* (See also ME, 126 – 7.) What he insists on there is that dreams do involve both memory and perception, because they occur when there is some perceptual stimulus to which, however, we associate the wrong memories. (See ME, 95, MM, 218 – 19. I will explain 'wrong' in a minute.) He is thinking of cases such as when we hear a door bang and dream of a gun-battle, and he quotes and also hypothesises some ingenious explanations along these lines, though one might think it would require considerable ingenuity to interpret all dreams in this way, even if it is not too far-fetched to suppose that *some* perceptual (including kinaesthetic, digestive, etc.) stimulus is involved whenever we dream. Of course we do not normally dream of incidents we can remember, but we do in some sense rely on memory, i.e. on past experience, for the material of our dreams; we do not, for example, dream of a colour we have never seen.

If we accept all this, however, there is left at best an awkwardness in Bergson's account when he treats the plane of dream as apparently synonymous with the plane of pure memory and at the opposite extreme from the plane of action; for, as we said before, pure memory is not conscious but dreaming is. Perhaps we could say that dreaming is as near as we can get to pure memory while remaining in some sense conscious; this should become clearer when we come to the relations between perception and action.

But first there is a point to note about Bergson's use of contractions. This is evidently a variable notion, in that the mind can adopt a greater or lesser degree of tension (e.g. MM, 129). We can live at a faster, more relaxed, rate, as when dreaming, or at a slower, more intense, rate. Why then do we not see red things as, say, blue when we are living intensely and so gathering greater swathes of objective vibrations into a single experience? For the minimum detectable period should now be greater than the 0.002" we referred to, and so should contain more than 8×10^{11} vibrations. Perhaps Bergson had in mind an answer to this question, but he does not seem to have given it – though sometimes he speaks as though each rhythm were confined to an animal species (M, 99).

6. *Perception and action*

Returning now to perception and action, it is here above all that Bergson is a pragmatist. Perception is not creative but selective. It selects from the totality of images those which are going to be useful for action, and we could not act unless we discriminated and focused our attention on only some of the aspects of things. (See MM, 304, 46.) He often talks of memory or the past as 'inserting itself' into the present and 'borrowing the vitality' of a present sensation i.e. becoming conscious (e.g. MM, 324, 320). As we saw earlier, we could never acquire knowledge of the past or a concept of it by starting with a present sensation or image and somehow referring it to the past: 'we shall never reach the past unless we frankly place ourselves within it' (MM, 173). (Cf. MM, 319.) Memory must, as it were, take the initiative and press itself forward to be realised in consciousness. It is the function of the brain in some not very clearly specified way to act as a filter and allow only those memories which are going to expedite our intercourse with the world to come through and influence perception in the ways described above. It is in this sense that in dreams the 'wrong' memories get through.

Bergson is plainly right in saying we cannot generate a concept of the past by starting from present sensations, just as we saw earlier that he rightly rejected the possibility of reaching the external world in that way. In both these respects he refuted in advance a good deal of philosophy written after he wrote. He is also right in emphasising how perception is dominated by our background. Elsewhere he shows how reading and hearing speech is affected by expectations, and it is equally true that it is dominated by interests: we see what we need to see (MM, 126, 134 – 6). No-one supposes, for instance, that an insect in any meaningful sense 'sees' the distant landscape, though its eyes may be as well fitted for doing so as ours are for seeing the distant heavens. But does not Bergson wildly exaggerate in treating *all* perception as a function of needs and possible action in this way? Granted that human needs range widely, over the intellectual, aesthetic, etc., do we not perceive many things in which we have no interest and which suggest no possible action to us? Did not primitive man perceive the stars long before he found a use for them in navigation or time-keeping, and did he not find these uses for them because he could already perceive them?

Perhaps Bergson means not that perception is defined in terms of action but that we only perceive what we can act on, namely the present, while our memories are of the past? It is not clear that this is enough for him. 'The *actuality* of our perception thus lies in its *activity*', while 'the past is essentially *that which acts no longer*' (MM, 74). (The italics are his; we could perhaps avoid sheer circularity by giving extra emphasis to 'acts'.) Only by emphasising the connexion with action as against mere speculation, he thinks, can we ensure that perception and memory differ in kind, not degree (MM, 74 – 5), though he insists that sensations too differ from pure memory in kind, not degree, this time because of their link with the body (MM, 179).

There seems in fact to be a multiple ambivalence in Bergson's thought about perception and consciousness. First concerning the world itself: his dominant view, I think, is that it is objectively real in its own right, albeit consisting of images in the way I tried to explain in chapter 4. Its division into bodies with absolutely determined outlines is artificial, but it does have 'natural articulations' which it is the business of science to rediscover (MM, 259 – 60). On the other hand sometimes he seems to go further and say 'the divisibility of matter is *entirely* relative to our action thereon' (MM, 292, my italics). But if the articulation of the world is something entirely contributed by us, and is the work of perception, which, however, subserves the purpose of action, what 'purpose' could action have unless there is already an objective world there? Maybe we see what we need to see – but why do we need to see it? This argument might be countered: we might be free to invent any world we like, but some worlds, once invented, would be such that we could act in them and others not. But, if we can do this, why don't we invent a world in which we don't need to act because our purposes are already accomplished? Furthermore what about the brain that is supposed to select the perceptual images needed for action? Presumably it, at least, must exist independently.

That is one ambiguity. Another concerns consciousness. When he is attacking the idea that memories are a kind of sensation, differing from perceptions simply by being less intense, Bergson insists that memories can be unconscious, and that consciousness is not the essential property of psychical states. He supports this by calling consciousness 'the characteristic note of the *present* . . . of the *active*', which is not 'the synonym of existence, but only of real action or of immediate efficacy' (MM, 181 – it is two pages after this that he talks of 'a non-perceived material object' as 'a kind of unconscious mental state'). Consciousness involves activity then and it is perception that brings memories into consciousness, though itself requiring memory if it is to be conscious because it involves duration. This suggests that consciousness will only exist in the context of a material world, and perhaps only when that world acts as a foil to us by providing resistance, for when things go smoothly we act automatically and 'unconsciously'; it is when we meet resistance that we 'come to' and are conscious of what we are doing – or at least this seems a Bergsonesque way of looking at it. But elsewhere, and especially in his later writings, where he defends immortality, he clearly does not wish to circumscribe consciousness so narrowly. He gives a hint in saying,

> Whatever idea we may frame of consciouness in itself, such as it would be if it could work untrammelled, we cannot deny that, in a being which has bodily functions, the chief office of consciousness is to preside over action and to enlighten choice. (MM, 182)

He does not develop this moderated version of his position, which does not fit all that happily with what he has just been saying, but before we take this any further there is another question to be mentioned.

7. The reality of the past

One of the strangest and seemingly most perverse of Bergson's doctrines is that of the reality of the past. Just how paradoxical a position he really wants to hold is unclear. We can all agree that in one sense the past is real in that it really did happen. The horrors of World War Two are real and not imaginary and it is often relevant to say just that. The trouble comes when we talk as though the past is somehow present, and the temptation to do this, which Bergson does not escape, is shown by the fact that not just in English the present and timeless verb-forms coincide; 'are real' a few lines above is ambiguous.

We have seen already how Bergson was impressed by Augustine's point about the squeezing out of the present, and if the present is squeezed out what is left but the past, or at any rate it and the future? We also saw how the role of duration in mental processes like uttering a word or sentence leads him to say that 'you cannot draw a line between the past and the present' (ME, 55). He concludes that bit of the discussion by saying, 'I believe that our whole psychical existence is something just like this single sentence . . . I believe that our whole past still exists. It exists subconsciously', and consciousness 'has but to remove an obstacle, to withdraw a veil, in order that all that it contains, all in fact that it actually is, may be revealed' (ME, 56 – 7). Whether

Bergson infers here that because any given moment in our life must be part of some specious present therefore all moments must be part of the same specious present, I am uncertain. But clearly he is not just calling the past real as opposed to imaginary. The idea that he wants the past to *survive*, with all the temporality that word implies, is again suggested when he remarks that, once we realise that consciousness is practical rather than speculative, 'there will no longer be any more reason to say that the past effaces itself as soon as perceived, than there is to suppose that material objects cease to exist when we cease to perceive them' (MM, 182). After suggesting that the spatially absent and the past both exist in the unconscious ('*the unconscious* plays in each case a similar part' (MM, 187)) he develops the point in a passage reminiscent, in a distorted way, of Kant's example of observing the house and the ship in the Second Analogy: we wrongly think of unperceived objects as more real than the past because we think of the objects as occurring in their correct spatial order; this we do because we have to take account of that order in acting on them, whereas our memories of the past, which we cannot act on, tend to occur in a random order (MM, 187 – 9).

But how about the doctrine that actuality is activity, that the past is what no longer acts (MM, 74)? As a definition this would be hopeless, for if we replaced 'no longer' by 'does not' we should not have distinguished the past from the future and the imaginary. But perhaps Bergson means something else. He sometimes gives the impression that the past could come out of hiding, as it were, and start acting, thereby becoming present; pure memories seem to do so when they become realised as memory images. But time and again he rejects the question *where* the pure memories exist in the meantime. Far from their being stored in the brain, it is the brain that keeps them at bay during waking life and lets through only those that will contribute to perception and thereby to action. The memories are not to be thought of as atomistic sensations which are 'there' waiting to come on to the stage. Nevertheless they are real, as we saw in the last paragraph. What all this suggests is that the pure memories do not exist now as entities – they exist (timeless present) in the past, but they have causal effects now, in so far as they generate memory images, which are present phenomena. They live in their effects, as a man might be said to live in his children or achievements, and this is a reality which is manifest now and does not belong to either the future or the imaginary. The sense in which all our memories survive would be that it always might be that an experience we had in the past but have apparently forgotten should at some future time come to mind and contribute to a perception. Writing to William James in 1903 he suggests that while (pure) memories are not 'things' they might all come to life as memory images in a sufficiently dream-like state of lowered tension (M, 588 – 9).

How far this somewhat anodyne account really represents what Bergson meant I am far from certain. It is perhaps suggested by a passage referring to the past as real because it conditions our present state and reveals itself in our character (MM, 191). It is certainly not free from difficulties. If a long past memory does have a causal effect now, does it have it by leaving a continuous trace in the brain (something Bergson would certainly reject)? Or by starting a continuous train of memory images leading up to the present (but Bergson evidently wants to cover cases where no such train exists consciously, and for them to exist unconsciously merely reinstates the problem)? Or by acting across a time-gap (not impossible perhaps but uncomfortable)? In his 1911 lectures on 'The perception of change' he attributes the preservation of the past to the indivisibility of change (CM, 183 (155)). This suggests the point about the unity of duration, the past leading imperceptibly into the present, which will only account for the role of the distant past if we appeal to the argument I mentioned earlier that if every moment of our life must belong to a specious present they must all belong to the same one. Does Bergson simply confuse knowing and known? Obviously our knowledge of the past is present, but the past isn't. But we need not accuse him of simply *confusing* the two. (Cf. H.W. Carr's (different) defence of him in Carr 1914.) Just as we cannot perceive the external world by just having sense-data, so we cannot know the past by just having memory-images. Just as the chair I see plays a real part in the perceptual situation without being here (inside me), so the past plays a real part in the memory situation without being now. Perhaps we could put it like this: he doesn't confuse the two senses of 'real' but uses the Augustine point to upgrade the non-temporal sense since the temporal sense has vanishingly small application. At the same time, though without emphasising it in this context, he treats past and future asymmetrically because of their different causal roles. We cannot know or be influenced by the future. At this point there seem to be hints of a causal theory of time, though they are not developed as such.

8. Mind and body

Finally we come back to the question with which this chapter began: the relations between mind and body. We have seen ample evidence that Bergson's position is not a simple one. One of his main targets is psychophysical parallelism, a view which he interprets sometimes as simply postulating a one-one correspondence between psychic and cerebral states, but sometimes as saying that the latter produce the former; an example of this second view comes in the formal definition he gives at the start of his 1904 paper on 'Brain and thought': 'Given a cerebral state, there will ensue a definite psychic state' (ME, 189), so that the contents of consciousness could be read off from a description of the brain. Not surprisingly this is associated with epiphenomenalism (MM, xiv). We have already met it under the guise of the brain-storage theory of memories in *Matter and Memory*, where he brings empirical arguments against it based on amnesia and aphasia, arguments, however, which he thinks only work directly in the case of memory, not in that of perception (MM, 83 – 5, 313 – 16). Despite saying in 1901 that neither experiment nor reason could *prove* that parallelism is impossible (M, 479), in 'Brain and thought' he mounted an a priori attack, claiming that the theory was inconsistent because it could only be stated by confusing, or oscillating between, two languages, those of idealism and realism (in one sense of those terms: ME, 193). The argument, which is reminiscent of one we discussed near the beginning of chapter 4 from MM, 9 – 12, makes some unclear use of words like 'present', 'portray', and 'equivalent' (ME, 196, 198 – 9), but seems to amount to this: if all we know of external objects is ideas, which would be the same even if the objects did not exist, and if the brain, which generates these ideas, is itself an object, and so known to us only as an idea, then this idea will coincide with all the ideas together, which is absurd. This contradiction is only avoided by inconsistently taking a realist view of the brain, helped by a distinction between perception and memory which, however, destroys the intended parallelism. On the other hand the realist thinks his ideas are caused by external objects, but also, if he is a parallelist, thinks they are caused (or possibly 'occasioned', to allow for occasionalism) by the brain, which he inconsistently still treats as an external object: he can attain consistency only by switching to an idealist view of the brain itself along with other objects.

This argument, if I have represented it correctly, is not very impressive. Bergson seems to assume that any causation of experiences by the brain must be all or nothing. But why could not the brain and the outer world be jointly responsible, the outer world making its contribution either directly or by itself causing brain-states? He seems to think his opponents inconsistent if they say this, because if the causation comes entirely from, or via, the brain we cannot know of anything outside the brain. If this were so as he rightly pointed out (MM, 9 – 12), we could not know the brain itself, as an object among others; we could not have our present concept of *the brain.* But is it so? Let us return to what I described as the fallacious step in the argument at MM, 9 – 12. Putting the argument as sympathetically as I could, I pictured him as asking how we could know of external objects 'if we have no access to them except via our brains' (p.91). The words are of course mine, but I suspect that the ambiguity in 'access' may represent Bergson's thought; does it mean 'if we cannot know about them except by somehow inferring their existence from what we know about our brains', or 'if we cannot stand in causal relations to them except through our brains'? It is not clear that the argument remains cogent on the latter interpretation. Such a causal theory of perception may be hard to elaborate in detail, but is not a non-starter.

Bergson's own view is that there is a one-many, but not a one-one, relation between cerebral and psychic phenomena; given a psychical fact we can infer a cerebral state but not vice versa (ME, 191, 208). Sometimes he gives the impression that the brain is a necessary condition for mentality (ME, 208), but elsewhere, especially in his later writings, he insists that only a small part of our mental life depends on the brain, so that survival of death is possible and indeed probable (ME, 58 – 9, CM, 52 – 3 (46)). In *Matter and Memory* he insists that the mind/body distinction must be made in terms of time rather than space. This idea follows on from his refusal to ask *where* memories are stored (MM, 191). The negative part of what he is saying is easier to appreciate than the positive. Even if memories were correlatable in a one-one fashion with parts of the brain we would hardly say, unless we are identity theorists, that the memories themselves have a geographical location, still less that the mental can be distinguished from the cerebral by being in a *different* place from it. But how does time come in? Consciousness involves duration, but so does the world from *Matter and Memory* onwards, so the distinction must be more complex. Both memory and perception involve duration but perception only involves it when supplemented by memory; pure perception, which 'is really part of matter' (MM, 297), is an ideal limit and instantaneous (MM, 325) in a way that is not equivalent to anything we can say about pure memory. So duration might in this sense distinguish memory from perception, and to that extent mind from matter. What mind or spirit does is 'bind together the successive moments of the duration of things' (MM, 295).

But all the time Bergson emphasises that the matter/spirit distinction is one of degree (MM, 295 – 6; cf. M, 957). The relations between mind and matter are discussed from a more general point of view in *Creative Evolution*, and he surmises in *Mind-Energy* that they have a common origin (ME, 18). But two questions arise for us at the moment, those of personal identity and immortality.

Personal identity is a topic on which Bergson says little, and some of that little is surprising, as when he says in a 1911 lecture on immortality that multiplicity and unity are concepts that apply to matter rather than mind, and that human souls are less distinct from one another than we might think (M, 959). An earlier lecture in the same course tells us that their substantiality can become a fact of experience, though apparently only for a Bergsonian philosopher (M, 947). Memory is always closely linked to personality for Bergson, and so it is no surprise to be told, in a discussion of dissociated personalities in 1916, that to speak of two personalities is to speak of two beings whose memories are totally different and absolutely independent (M, 1227). But he is concerned to use the point rather than discuss it, and does not pursue it further. Earlier in the same lecture, however, he distinguishes two problems, a metaphysical one which he thinks artificial and a psychological one which he takes seriously. The metaphysical problem asks what unites the different states of consciousness into one, and he thinks that philosophers from Plotinus or even earlier to Kant have answered it by postulating two souls, one in time and one out of it. But instead of asking with Hume (whom he does not mention here) what unites our perceptions at a time he concentrates on what unites them over time. As we would expect, he thinks the question depends on a false presupposition, that conscious states occur in a discrete 'cinematographic' series. This ignores everything he has said about life and duration, and the truth is that 'what characterises the person is in our view the *continuity of movement* of its inner life', as he expressed it in the summary of his 1914 Gifford lectures on the same theme (M, 1070), a continuity which requires effort (M, 1225) and which animals do not have – though he adds, strangely, that they *may* have speech and intellectual abilities (M, 1070). The second problem in the 1916 lecture concerns things like multiple personality, but the emphasis is psychological and not metaphysical, except negatively; he seeks to explain the relevant states without postulating more than one person and then to ask how they arise (M, 1224ff.).

What would Bergson say about total amnesia? Presumably that it never occurs, except at the conscious level, because all memories are preserved. We have already seen the difficulties this involves once memories are dissociated from the body. The body, and in particular the brain, acts as a filter which lets through into consciousness just those memories which are useful for action, inhibiting the rest so that we do not get swamped and effectively paralysed. (Cf. e.g. ME, 7 – 9, M, 1212.) At one point at least Bergson goes further. According to notes made at a lecture-course he gave in 1910 – 11 he makes the brain the principle of individuation, saying that without it there could be no distinct and countable personalities (M, 870). Given his view that only a small part of our mental life depends on the brain this illustrates his rather casual attitude to what we might consider the fundamental concept of a person, and it is not surprising that in 1916 he accepts that societies can be persons, apparently in the full literal sense (M, 1232); he says a bit later that he merely cites this thesis without defending it (M, 1233), but he has just said he accepts it, and he credits the opposing thesis, which involves that nations have no moral rights, to 'certain recent German theoreticians' following Hegel; in 1916 he did not approve of Germans.

Matter and Memory on the other hand, referring to the practical role of the body, talks of our needs as 'carving out' a body from the sensible continuum and distinguishing it from other bodies (MM, 262). This suggests that we are prior to our bodies, and the very passage I have just referred to as individuating via the body talks of the body as 'forcing personalities to dissociate and distinguish themselves' (M, 870). The proper conclusion seems to be that Bergson really thinks of individual selves as some sort of crystallisation out of the general life-force which occupies him so greatly in *Creative Evolution*, where again we have this rather casual attitude to personal identity (CE, 271 – 2) and to the personality of societies (CE, 273); he explicitly makes matter the principle of individuation (CE, 284).

But what about survival and immortality? He is more certain of the former than of the latter, which may fit with what I have just been saying. (See e.g. M, 988.) But there are many difficulties. He succeeds no better than most dualists in saying *how* mind and body are related, or how the body and brain fulfil the pragmatic roles he gives them. Our needs carve out a body – but how do we have the needs unless we already have a body? (Cf. T.A. Goudge 1967: 290.) A big question is whether we can act without a body, for 'consciousness is synonymous with choice' (ME, 11) and so in effect with action. It is not 'the synonym of existence' (MM, 181), so there could be survival in a state of permanent unconsciousness – but that would hardly satisfy Bergson or anyone else. On the other hand the role of the brain is to suppress memories when their proliferation threatens to swamp action – though how it does this suppressing is never explained – so that with the brain no longer operating one might expect a total memory of the past (M, 1213), which must mean a conscious memory, since we have an unconscious memory of it anyway. Kolakowski suggests the afterlife might resemble dream perception. (Kolakowski 1985: 49; cf. also MM, 322.) Verbally these positions might be reconciled by saying that '*in a being which has bodily functions*, the chief office of consciousness is to preside over action and to enlighten choice' (MM, 182, my italics), so that consciousness could have some other role in its 'untrammelled' state – but we are never told what role, nor why, if brain damage can cause unconsciousness, as he could hardly deny, brain destruction should cause consciousness. He says 'the only reason we can have for believing in the extinction of consciousness at death is that we see the *body* become disorganized' (ME, 59, my italics). But what else could we 'see'?

All in all then it does not seem that Bergson has any very satisfactory answers to questions about personal identity, the relation of mind to body, and the possibility of survival. But no doubt it could be said on his behalf that he is in good company.

VI

Epistemology

1. Introduction

In 1903 Bergson published an article called 'Introduction to metaphysics', which despite its title is at least as much an introduction to his epistemology; he reprinted it in the collection *The Creative Mind*, described in its preface as containing material 'relating this time to the task of research', as opposed to the earlier volume *Mind-Energy* which contained material 'dealing with the results of some of my work'. It is here that he first discusses at length the notion of 'intuition' which he contrasts with 'intellect' or 'intelligence', but the situation is complicated by a third term, 'instinct', which is related to both the others but especially to intuition, giving rise to Russell's notorious gibe about intuition being at its best in 'ants, bees, and Bergson' (Russell 1914: 3). All these notions are developed further in Bergson's fourth book, *Creative Evolution*, as well as in sundry other items in *The Creative Mind*, and it is in *Creative Evolution* in particular that he deals with instinct (especially pp.141ff.). Of these three notions it is clearly intelligence and intuition that are the most important, but let us start by looking at intelligence and instinct, both of which seem to come before intuition in the course of evolution.

2.*Intelligence and instinct*

What strikes one at the outset as paradoxical about Bergson's approach is the way in which he seems to put intelligence and instinct on the same level. Surely intelligence is much 'higher' than mere instinct? Bergson does not explicitly deny this, though some of what he says might suggest a denial. The judgement will obviously depend on the standpoint from which it is made, and here intelligence has an advantage since instinct does not make 'judgements' of that sort at all. But even intelligence might admit that sometimes and in some sense instinct has the advantage. We think more highly in some circumstances of a person who acts from maternal instinct than of one who acts from cool calculation – we think more highly of them 'as a person', and may still do so even if the calculation was not egoistic but utilitarian.

Such cases perhaps are rather special. The main criterion, from Bergson's point of view, would be a practical one: is intelligence or instinct of greater advantage to its possessor – or perhaps to the species to which its possessor belongs, since Bergson, as we have seen before and shall see again, is not always as interested in the individual as we might think he should be?

A main feature of Bergson's approach is that he treats intelligence and instinct not as succeeding one another in time but as diverging from a common source, or representing divergent and alternative tendencies in the course of evolution, though never completely isolated from each other. (See CE, 142, 143, 149.) Both of them are practical in nature at least in origin, aiming at action rather than knowledge, though of course knowledge may be needed as a means; see CE, 163, in particular.

Instinct and intelligence are 'divergent solutions, equally fitting, of one and the same problem' (CE, 150), 'one assured of immediate success, but limited in its effects; the other hazardous but whose conquests, if it should reach independence, might be extended indefinitely' (ibid.). This sounds promising as a way of distinguishing the roles of the two in the economy of nature, but it is not clear why the distinction should exist in that form. Intelligence is clearly self-developing and cumulative in a way that instinct is not, a fact connected with the notorious unadaptability of instinct; but instinct does exist and has developed as far as it has. What, in Bergson's own terms, is to prevent it developing ever further? After all it is not, for him, an ossified intelligence, which would presuppose an intelligence to ossify. (See CE, 183 – 4, 185 – 6, 178, 179.) He makes great play of the extreme complexity and apparent adaptability that instinct can reach, as when wasps sting their prey in just the right way to cause paralysis without death and so provide a continuing food supply for their larvae (CE, 180 – 3); but, however this instinct developed, since it did develop why should it not go on doing so? This, however, is not a fundamental criticism; it merely spoils the symmetry rather. He could reply that intelligence is better placed to develop in a systematic and reliable fashion.

A more serious criticism is whether on Bergson's view intelligence is not tacitly presupposed somewhere along the line. For how does the wasp operate? It 'knows that the cricket has three nerve-centres . . . or at least it acts as if it knew this' (CE, 182). The wasp is not an entomologist. What it does have is 'a *sympathy* (in the etymological sense of the word) . . . which teaches it from within, so to say, concerning the vulnerability of the caterpillar' (CE, 183). This feeling, we are told, may result not from outward perception but from the mere presence together of wasp and caterpillar, expressing 'the *relation* of the one to the other' (ibid.). In fact, 'Instinct is sympathy' (CE, 186). If this is not mere sympathetic magic (and, if it is, why does it not occur in all sorts of other comparable cases?), we need to be told a bit more. What is the cash value of the 'as if' and 'so to say'? The wasp can hardly 'know' where the cricket is vulnerable simply by knowing where it is itself vulnerable, as I might instinctively hit a man on the nose, since presumably it does not have the same structure as a cricket. What in fact is the 'relation' between them that is in question? We shall not come until the next chapter to the force that operates the evolutionary process, but does it not look as if that force itself will be having to exercise intelligence in making the wasp develop as it does – unless it too operates by some sort of 'sympathy', which would merely put the problem back a stage? Bergson in fact thinks that science will probably never be able to analyse instinct completely so long as it uses its present methods of explanation, i.e., presumably, so long as it keeps to the methods of intelligence, which amounts to saying, so long as it remains science rather than metaphysics. (See CE, 177, and §§4ff. below.) He adds that some sort of racial memory is involved, for 'the instinctive knowledge which one species possesses of another on a certain particular point has its root in the very unity of life' (CE, 176) and involves seemingly forgotten recollections. This is presumably intended to get over the point about the wasp and the cricket having different structures by bringing in their common origin; but it is hard to see how this will help, since the wasp has not developed from being a cricket, and the cricket too has developed since they parted company, and what is more, as Bergson insists in other contexts, along unforeseeable lines.

What then is instinct? We have just seen that it cannot be analysed entirely in scientific terms. But here I think we meet a certain ambiguity that pervades much of Bergson's thought in this area. Certainly, he agrees, we must not simply 'stop short before instinct as before an unfathomable mystery' (CE, 184). But, since intelligence and instinct represent divergent evolutionary paths,

'Why, then, should instinct be resolvable into intelligent elements? Why, even, into terms entirely intelligible?' (ibid.) The ambiguity lurking here is this: does it follow that because instinct is not itself an intellectual faculty therefore no account can be given of it in intelligible terms? Obviously this does not follow in general or we could never understand anything that did not itself involve understanding. Is instinct a special case?

Certainly there is something mysterious about instinct, even if not unfathomably so. In its uncanny way of letting its possessor know what is not accessible by ordinary means of knowing it resembles action at a distance, something Bergson expresses by saying it 'has the same relation to intelligence that vision has to touch' (CE, 177). In the cases we are considering, where instinct involves a potential or unconscious innate knowledge (CE, 158), instinct seems to operate directly where intelligence operates indirectly. Instinct uses already organised matter (the body and its parts) while intelligence constructs its own instruments from unorganised matter; this gives intelligence the advantage of adaptability, which serves it well in its later developments, but at the outset it relies upon instinct, in the sense that creatures capable of using intelligence could only have evolved 'on the wings of instinct' (CE, 150). Presumably the point is that ancestors of the creature in question must have relied upon instinct; but what is not quite clear is whether instinct plays a role in the process of evolution itself; after refusing to draw a sharp line between 'the instinct of the animal and the organizing work of living matter' Bergson says that 'instinct perfected is a faculty of using *and even of constructing* organised [i.e. organic] instruments' (CE, 147). (Bergson italicises the whole sentence.) At any rate the directness of instinct will consist in its owner's innate knowledge of how to use its organised instruments and perhaps how to evolve its own body or those of its descendants, and because this concerns a particular given body Bergson describes it as knowledge of matter or things while intelligence is knowledge of form or relations, because it can be flexibly applied to any suitable matter that presents itself and is not limited to something given (CE, 155 – 7). This flexibility, which also allows it to speculate, gives intelligence an obvious advantage, and he sums up the situation by saying 'There are things that intelligence alone is able to seek, but which, by itself, it will never find. These things instinct alone could find; but it will never seek them' (CE, 159).

The problem of the nature of instinct cannot be separated from that of its origin. Instincts, or some of them, seem to embody a sort of magic knowledge that could not have been acquired in the way one would expect for such knowledge. But a suitable account of the origin of such instincts might show that such knowledge was only illusory. The creature was indeed behaving 'as if' it knew all sorts of facts about, for example, what would happen to its larvae if it laid them in a certain place, and also desired that that should happen. What is needed, if we are to part company with Bergson, is an account which will explain the series of events without needing a relation we should call knowledge, in any but a metaphorical sense; an analogy would be an account of the behaviour of homing missiles, though there of course knowledge plays a part in the origin of the behaviour, if not in its nature. The puzzle is to explain why the insect lays its larvae in a place which is beneficial to them. The solution for the non-Bergsonian will be to find some other feature, F, such that we can explain why the larvae are laid in F-like places and also explain why being F-like is correlated with being beneficial.

The solution in any given case will require detailed scientific investigation, which Bergson, like Russell, was not averse to. (See his 1916 Madrid lecture (M, 1197 – 8), and also ME, 37 – 8.) If he says that no solution can be found (CE, 177) he will be vulnerable to the advances of science. He may reply that this puts him in no worse position than any controversialist, since he agrees that philosophy is 'rigorously based on experience (internal and external)'; even a tentative solution is better than none. (See his 1912 letter at M, 964.) This reply will only carry conviction if his own solution is indeed a coherent one. It involves what he himself calls 'innate knowledge (potential or unconscious, it is true)' (CE, 158), and knowledge that is qualified by 'as if' (CE, 182). Now knowledge, we might say, has two features in particular: it is associated with consciousness, and it requires a certain connexion, of an admittedly controversial kind, between knower and known.

Consciousness is linked with both instinct and intelligence (CE, 150 – 3). It is the 'background' from which instinct and intelligence 'stand out' (CE, 196). We would not normally attribute knowledge, except by analogy or metaphor, to anything that was never conscious, and neither does Bergson; stones have neither knowledge nor instinct. But something can now have knowledge, and even now exercise knowledge, though it is not now conscious. We all do 'unconsciously' lots of things that require knowledge. It might be said that we must be conscious of something in these cases, if not of the thing we know; if I 'unconsciously' avoid obstacles while walking deep in thought, this is only because my consciousness is taken up by the thought so that there is none left for the obstacles. If I were under a deep anaesthetic I would not avoid them at all, even if my legs somehow propelled me along the road. True, I might be somnambulating; but I will not get far if I somnambulate with my eyes shut, and consciousness in fact is only just beneath the surface: as Bergson points out, I will only remain unconscious as long as things go smoothly; if my action is frustrated in any way consciousness will instantly return. One is inclined to say that consciousness here is a matter of degree. Bergson indeed says that the somnambulist's unconsciousness may be 'absolute' (CE, 151); but it is not clear what case he is thinking of. Does the somnambulist avoid obstacles because his eyes are open and he has some sort of 'subliminal' perception of them, or not? The term 'absolute' suggests the latter; but then, unless we are to posit a sheer miracle, we shall have to say he avoids obstacles because he is walking a route already familiar to him. He could do that by a sort of pre-programming without any *current* stimulation from the environment. But that simply takes us back to previous occasions when the programming occurred. Must he not then have been conscious? Well, cannot conditioning occur in machines as well as in men? But it is in just those cases that we do not talk of knowledge, except in inverted commas, as with the missile.

So knowledge involves consciousness, whether or not its possessor has to be conscious all the time he is exercising his knowledge. Bergson agrees. But it is instinct we are trying to explain, and whatever by way of knowledge it may involve. It will not do for consciousness to be a mere accidental accompaniment of such knowledge, for what would be the point of insisting on its presence in that case? Consciousness must have a role to play, as it obviously does where we acquire knowledge through perception.

But this brings us to the second feature of knowledge, the connexion that must exist between knower and known. It is tempting to talk of justification here, but dangerous if this implies some third element or added process. I know where my legs are, and what I am thinking about, and whether I am enjoying my lunch, and if asked to justify my claims to know such things I could only say that I just do. All the examples I have given, however, have in common something that might be crudely but convincingly described by saying that there is no gap between knower and known. The known is in each case a state of myself – though of course not all states of myself could be known in this way. But what our wasp has to know is not something about itself but some very complicated facts about a cricket – and even more complications are involved in another example Bergson gives, that of the Sitaris, which he vividly describes at CE, 154 – 5. Can we really say that the connexion between knower and known here is so close that there is no gap between them that requires an epistemological bridge? Bergson himself is well enough aware of the difficulty and insists that 'The knowledge, if knowledge there be, is only implicit' (CE, 154). But now what has happened to the role of consciousness? If the insect is not *now* conscious of all these things it needs to know, then was there some previous time at which consciousness helped it to get this knowledge? Presumably not. Bergson in fact is in a dilemma. Either what the insect has *is* knowledge, without any 'as if' about it, or it isn't. If it is, then where is the connexion between knower and known that we have demanded, and what is the role for consciousness that he himself agrees in demanding? If it is not, then what is the solution he is offering to the original problem? Are we reduced to saying that it is 'as if' he has offered a solution?

3. Consciousness: its two senses

Are we being unfair on Bergson and forgetting what he actually says about consciousness? This is that it arises when things are not going smoothly, when there is a choice to be made. 'It measures the interval between representation and action' (CE, 152) and though, as we saw, it provides the 'background' for both instinct and intelligence it is much more closely linked with intelligence, because only there do hesitation and choice arise (CE, 152 – 3). But the link with consciousness is still there. Not only does he insist that both instinct and intelligence involve knowledge (CE, 153), but the unconsciousness of the somnambulist is distinguished from that of the stone; in the latter consciousness is absent, in the former merely 'nullified' and ready to reappear at any moment (CE, 151).

But what is consciousness, and does Bergson always say the same thing about it? We have seen in this and the last chapter something of his attitude towards it, but there is a further ambiguity to be considered. Consciousness seems at first to be something clear and definite, which distinguishes men and animals from plants and minerals, give or take a few doubtful cases. What encourages this and makes it seem so important is its moral significance. A large slice of morality concerns the infliction of pain, in one form or another, and I have never heard of a Vegetable Liberation Front. (Conservation policies normally serve *human* interests.) But there are notorious difficulties. Consciousness cannot exist on its own. We cannot be *just* conscious without engaging in some conscious activity or experience. Nor is this quite like saying something cannot be just coloured without being red or blue etc. – or at least to say it is would be a substantive claim needing a lot of defence. The modes of consciousness do not seem to be related together in the simple way that modes of colour are, and consciousness does not seem to be a determinable in the simple way that colour is. This seems to be at least partly because there are relatively simple links between being coloured and such properties as being extended, located, and detectable by vision, despite problems about transparency or the visibility of stars subtending no detectable angle at the eye. There seem to be no comparably simple links to other properties in the case of consciousness.

But there is another sense of 'conscious' where it means something like 'self-conscious' or 'reflective'. That this is another sense is clear because animals we would all (except Descartes) agree to be conscious in the first sense we would also agree are never conscious in the second: they never act consciously of what they are doing and are never conscious of themselves as such. This is a narrower sense than the first, which it presupposes; every being and every act or state which is conscious in the second sense is conscious in the first but not vice versa. Both senses are complex and both are matters of degree, though if there is consciousness in either sense there must be something that *is* conscious. If there is pain in a tree there must be some answer to the question whether it is the leaf or the branch or the tree, or indeed the forest or the earth, that feels the pain; if I feel a pain in my leg my leg doesn't also feel a pain – or, if it does, that is a quite separate and contingent matter and of no concern to me. This, as we saw in the last chapter, is something Bergson is inclined to soft-pedal if not ignore.

If we now ask what Bergson says about all this, it seems that he does not really distinguish the two senses of consciousness. Perhaps he would want to claim that they do indeed amount to the same thing, but he never explicitly does so, I think. Usually he seems to have the second sense in mind, for it is this that is particularly likely to arise when activity is frustrated or a choice is called for. It is this that is more obviously associated with intelligence than with instinct. But what has this to do with a conscious experience like feeling pain? That is not a matter of choice or frustration. Of course we shall choose to avoid the pain and it may frustrate what we are doing. But that is irrelevant; *that* choice and frustration presuppose the pain and cannot serve to analyse it. If we are frustrated in any way this may cause us distress; but there are plenty of pains, both physical and mental, that do not presuppose any activity to be interrupted: what about pains that wake us from our sleep? Nor could we say that pain is the frustration of some physiological process, for not only is its connexion with any such process contingent, but what would justify the word 'frustration', unless we say, trivially, that what is frustrated is whatever process would occur without the pain? And why are lethal pains often much less severe than medically trivial ones? On the other hand Bergson seems to be thinking of the wider sense of consciousness when he speculates, as we saw in the last chapter, about 'consciousness in itself, such as it would be if it could work untrammelled' by the body instead of having to 'preside over action and to enlighten choice' (MM, 182). One would think this sense is again dominant when he says of man as against animals, 'He is not limited to *playing* his past life again; he *represents* and *dreams* it' (CE, 190; italics his). (Cf. his 1911 Huxley lecture: 'it [consciousness] means, before everything else, memory' (ME, 5).) But the position is far from straightforward. Dreaming seems far removed from action and reflection, yet it is precisely in men as opposed to animals that it occurs, at least primarily. In fact it is the association of consciousness with non-human animals and with life in general that suggests the wider sense of consciousness where it does not involve choice and reflection or consciousness of what one is doing, and yet it is in this same non-human sphere that consciousness is for Bergson linked most closely with action and so with choice. His strange view of the role of choice is illustrated when he suggests that certain unicellular organisms may have consciousness because they 'hesitate between the vegetable form and animality', as though they could not decide which way to go. Admittedly this case is marked by 'consciousness asleep and by insensibility' as against the 'sensibility and awakened consciousness' of the animal (CE, 118).

One might sum up Bergson's attitude to consciousness by saying that he thought in terms of the narrower reflective consciousness in so far as he associated consciousness with life in general, and that though he rightly refused to see consciousness as a thing in itself, a sort of custard poured over the pudding of life's activities, he went too far in the opposite direction in his linking of consciousness with choice and activity, and ignored the passive form that consciousness can take as well in things like feeling pain. Perhaps he really had in mind the idea that pain or painful consciousness can only occur in creatures that are capable of reacting to it by avoidance etc., and so not in plants; this gives a connexion between pain and action but not of the kind that he seems to want. The animal reacts *to* pain, which is presupposed, and no choice need be involved; the reaction can be quite automatic and the animal need have no choice about whether to react, or even how.

4.Instinct and intuition

Instinct and intelligence diverge from a common source. But, whereas intelligence does not lead on to anything beyond itself, instinct does, namely to intuition. This remark will need qualifying a little, but it is broadly true; instinct is superseded in a way that intelligence is not, and intuition replaces instinct as the main rival to intelligence. Bergson says comparatively little explicitly about the relations between instinct and intuition, but he does make clear that intuition is a development of instinct. A letter of 1915 says of 'human intuition' that it 'prolongs, develops and transposes into reflection that which remains of instinct in man' (M, 1150). The most explicit statement comes earlier: ' . . . by intuition I mean instinct that has become disinterested, self-conscious, capable of reflecting upon its object and of enlarging it indefinitely' (CE, 186). This seems to some extent to bring it nearer to intelligence, and a page later we read,

> though it [intuition] thereby transcends intelligence, it is from intelligence that has come the push that has made it rise to the point it has reached. Without intelligence, it would have remained in the form of instinct, riveted to the special object of its practical interest, and turned outward by it into movements of locomotion. (CE, 187)

This need for intelligence explains why intuition only occurs in man, and is the qualification I referred to a moment ago. Later the point is made the other way round, as it were, when we are told that in the course of evolution 'intuition could not go very far' and 'had to shrink into instinct', apparently because of the sort of body and brain it was associated with (CE, 192). None of this is affected by Bergson's repudiation elsewhere of 'the kind of person who would insist that my "intuition" was instinct or feeling. Not one line of what I have written could lend itself to such an interpretation . . . my intuition is reflection' (CM, 103 (88)). What he is repudiating is that intuition was simply 'a sort of relaxing of the mind' (ibid.) as an alternative to hard work, an interpretation no doubt congenial to many among the mass following his writings attracted.

The ways in which intuition goes beyond instinct are that it is conscious, confined to humans, and reflective. In these respects it is nearer to intelligence. Yet it is plainly different from intelligence and it is in the contrast between those two that Bergson is mainly interested. Instinct plays only a minor role, and indeed is really discussed only in *Creative Evolution*. What intuition has in common with instinct is that its grasp is of life, while that of intelligence is of unorganised matter. But while instinct and intelligence grasp their objects incompletely with intuition the grasp is more complete.

But one feels that what really allies intuition with instinct is that they both seem to involve knowledge without any apparent means of acquiring it. It is this after all that makes the two terms so often nearly synonymous in ordinary life. However, in the 1915 letter quoted earlier, on which the last paragraph was based, Bergson says that *practical* knowledge (his emphasis), whether it takes the form of instinct or intelligence, is absolute rather than relative provided it remains within its own sphere (life or matter respectively). When it tries to invoke the other sphere it becomes merely relative, and this is seen when knowledge of life claims to give us conceptual intelligence, as happens with mechanism, or when we represent matter by images drawn from the sphere of life, as happens with hylozoism. The term 'absolute' is linked at one point to and apparently glossed by 'from the inside' (M, 1149 – 50).

The letter effectively ends there and is not very easy to interpret. Some things will become clearer later, I hope, but the main issue that the passage seems to raise is this: are we faced with a basic epistemological distinction between two faculties, intelligence and instinct/intuition, or are the two symmetrical in the sense that they operate in parallel ways within their respective spheres, the distinction being one between the spheres (matter and life) rather than between the faculties themselves? This would be rather like saying that history and geography are parallel academic studies, distinguished by the fact that one concerns the disposition of things in time and the other their disposition in space. A further alternative would be that the distinction is indeed primarily between the objects of the two faculties, but that this distinction is itself such that it forces one of kind between the faculties, as though the sort of thing that history told us about the disposition of things in time was radically different from the sort of thing that geography told us about their disposition in space: is instinct/intuition just the sort of knowledge one can have of life, intelligence the sort one can have of matter?

It seems clear that the second of the three alternatives (that the distinction is simply between the objects) is not Bergson's view. What militates against the third view is that there is something called 'relative' knowledge as well as 'absolute' and this the faculties can have of each other's objects – though since it leads to mechanism in the one case (when intelligence tries to know life) and hylozoism in the other, and mechanism at least is something he firmly rejects, it is not clear whether relative knowledge is knowledge at all in any but an inverted commas sense; it certainly alters the nature of its object (M, 774). It might be premature to press this last point too far. Perhaps it is not inevitable that the use of intuition to study matter should lead to hylozoism, and though he refers to matter as 'that for which it was not made' (M, 1150) this perhaps need only mean that matter is not that for which it is primarily suited. However, we shall see shortly that he changed his view somewhat in this area.

So far we seem to be pressed towards the first of the three alternatives, that the epistemological distinction is basic and primary. One of the considerations that support this is the way Bergson insists that any account of his views will distort them if it does not unceasingly give first place to their central point, 'the intuition of duration' (M, 1148); and duration, as we saw when discussing this passage in chapter 2, belongs to the non-conscious world as well as to life from *Matter and Memory* onwards, though not before. Science as an intellectual study may be able to ignore the time intervals between events and treat them as simultaneous, but if the material world contained no duration the living world could hardly be embedded in it in the way it is. On the other hand he distinguishes philosophy from art because it studies matter and *therefore* appeals to intelligence, though intuition is its specific tool (M, 1148). He also adds that he conceived of his theory of intuition long after conceiving of that of duration, from which the former derives and without which it cannot be understood (M, 1148 – 9).

The situation is complicated by the change of view I referred to just above. Bergson refers to this in three footnotes added later to the 'Introduction to metaphysics'. (See CM, 305 – 6 notes 20, 26, 27 (Q, 1392, 1423, 1425).) Here he explains that after 'Introduction to metaphysics', and in particular in *Creative Mind*, chapters II and IV, as well as in *Creative Evolution*, he tightened his usage of 'intuition' and of the distinction between science and metaphysics. Originally intuition was the knowledge of mind by mind and secondarily mind's knowledge of what is essential in matter, as against intelligence's practical knowledge of it. This, he tells us, is how he used the term in 'Introduction to metaphysics'.

But later he used it more strictly and narrowly in keeping with his stricter separation of science and metaphysics, whereby science becomes limited to the study of matter, with intelligence as its method, and metaphysics to that of mind, with intuition as its method. (See also CM, 42 – 52 (37 – 45).) Here we have an emphasis on the different objects of the two methods; but this is nevertheless introduced by a distinction between the methods themselves: science employs measurement, while metaphysics aims to '*sympathise*' (his emphasis) with reality. (CM, 305 note 20 (Q, 1392).) The distinction between their objects is consequent on this.

All in all then it seems that the first alternative is the correct one, despite some remarks suggesting the third. Even if the idea of duration came first and is essential for understanding intuition, the tying down of intelligence and intuition to matter and mind as their respective objects comes later still, and the core of the distinction is that between measuring and 'sympathising'.

5. *Intelligence and intuition: preliminaries*

To turn now to the distinction between the faculties themselves, we have already seen that intuition 'transposes into reflection' what remains in man of instinct, and thus becomes theoretical rather than practical. In fact it seems that only instinct can take this path, though it requires intelligence to help it, so that an initial distinction between intuition and intelligence is that intuition is properly theoretical while intelligence is properly practical even though it can be theoretical, and it is *qua* theoretical that it helps instinct (CE, 168). But surely science and mathematics are eminently theoretical. Not necessarily, in Bergson's eyes. Science is concerned to help us manipulate the world, and mathematics, which we will return to a bit later, he is not always very good at distinguishing from science, as we saw when discussing necessity in chapter 3. Intelligence always takes the cinematographic approach, which we have discussed earlier and which is elaborated in *Creative Evolution*, chapter IV, in particular, and it does this because its function is to preside over actions, and actions are envisaged as leading up to results, which are states, and so immovable; and to perceive the result in this way intellect must similarly perceive its surrounding as immovable. If we did not look at matter in this way but saw it as flowing, 'we should assign no termination to any of our actions. We should feel each of them dissolve as fast as it was accomplished . . . ' (CE, 316). He illustrates this with the rather unfortunate example of raising one's arm, where one looks straight to the end without envisaging all the neural and muscular contractions this involves (CE, 315 – 16). He seems to treat these contractions as means to an end, though perhaps he means they would be means if we did envisage them consciously. But more serious is that our aim might be not to achieve the final posture but to be raising the arm, for example, as a form of exercise, and surely many of our aims involve activities rather than results, though of course we can aim at results. Are we to say that aiming at activities involves some other faculty than intellect? When I learnt to ride a bicycle my intellect told me, on the basis of bitter experience, to lean into a turn and not away from it if I wanted to continue the activity of riding. Bergson seems to draw the line in the wrong place. One might also wonder why practical ends should not require a knowledge of art, morality, other minds, to say nothing of duration, all of which on Bergson's terms are subjects for intuition, not intellect. (See Stewart 1911, chapter III.) In fact it is clear that intuition is not just given to us for speculative purposes, but is essential to life on the human plane at all, if it has anything like the role Bergson wants to give it. The most he could say is that intelligence is needed as well. Stewart also wonders why intelligence is so useful if it gives us such a distorted view of the world, the cinematographic view. Bergson would reply that we have to treat a moving thing as occupying a position at a moment if we are to act on it. But it is odd that our actions should turn out to be successful if, as he thinks, it is not true that the mobile is at a position at a moment, since it is in an indivisible state of motion. In fact it is hard to reconcile these three propositions: Intelligence has absolute knowledge of its own object, matter. (Cf. M, 756.) Intelligence views things cinematographically. The world embodies duration, which is indivisible. Perhaps we must emphasise the distinction between matter as an ideal limit and the world which contains duration and therefore spirit. (Cf. CM, 37 (33).)

But the basic distinction between intelligence and intuition is, as we saw, that between measuring and 'sympathising'. Measuring seems clear enough, but what does 'sympathising' amount to? We have seen something of it already when discussing duration in chapter 2. For indivisible duration cannot be measured, but must be apprehended as a whole, like a melody. Stewart (1911: 228) points out that in apprehending a melody we must apprehend its notes separately as well as in synthesis. This is true, but Bergson would still say that the notes are not parts of the melody in the sense of being a subset of a total set of parts which would exhaust the melody. This gives us a negative feature of intuition, that it is holistic and cannot involve the sort of synthesising that measurement yields.

Another area where measurement, at any rate in the extensive sense, seems out of place is in our apprehension of qualia as opposed to the scientific properties underlying them. Whether these are magnitudes in any sense we discussed in chapter 1, but we were not there concerned with intuition. Bergson constantly insists that we cannot approach duration from the outside; we must place ourselves within it at the outset. (See e.g. CE, 315.) The same might be said about qualia, in the sense that if we have not got (say) sight no amount of intellectual excogitation will give us the concept of red that a sighted man has – whether or not M. Tye is right in insisting during his defence of physicalism that the blind man who acquires sight need not be learning new facts but only new ways of knowing old facts (Tye 1986).

Both this case, if it counts as one of intuition, and that of duration raise a problem for the view that mind is the only object of intuition, for duration belongs to the world itself, as we saw above, and so do qualia, as we saw when discussing images at the beginning of chapter 4. Yet if knowledge of qualia does (*pace* Tye) count as knowledge of something new, and Bergson certainly seems to think knowledge of duration does, then to what *does* this knowledge belong, if not to intuition? It can hardly belong to intelligence as Bergson views it. When we perceive qualia we perceive, with a high degree of tension, 'trillions of oscillations' (CE, 317), and 'the qualities of matter are so many stable views that we take of its instability' (CE, 318). But this is plainly not the sort of experience to which we can apply measurement.

6. *Immediacy*

One theme running through this part of Bergson's philosophy is opposition to Kant's view that all our knowledge is fitted to the limitations of the human mind and that there is no such thing as knowing things as they are in themselves. There can, we saw, be knowledge, albeit limited in scope, which is 'absolute' in that it leaves its object intact while 'relative' knowledge alters its object; furthermore though this knowledge may be only partial its scope can be extended indefinitely (M, 774). Elsewhere we are assured that both science and metaphysics, and therefore both intelligence and intuition, can be absolute. (See CM, 42 (37).) What is not yet clear is how this notion of absoluteness relates to another important notion, that of immediacy. In a note on the term 'immediate' written in 1908 Bergson refers to both *Matter and Memory* and *Creative Evolution* as trying to refute the view that what is immediately given to consciousness is confined to the subjective; in particular, *Creative Evolution* claimed that immediate intuition grasps the essence of both life and matter. To say otherwise, he continues, is to deny the possibility of two different kinds of knowledge, 'one static, by concepts, where there is indeed separation between knower and known, the other dynamic, by immediate intuition, where the act of knowledge coincides with the act generating reality' (M, 773). Leaving aside the problems raised by the last claim, it is clear that the knowledge which uses concepts is intelligence. (Cf. e.g. CM, 198 (168).) The point is presumably that intelligence, which separates knower from known, is not immediate, while intuition is. But intelligence is or can be absolute. So being immediate is not the same as, and is not implied by, being absolute, and since knowledge is absolute when it does not distort its object, being immediate must mean something else.

Berthelot distinguishes two senses of 'immediate' (Berthelot 1913: 208 – 12, 317 – 18, 327). In one, used by the Scottish common-sense school, it means what is currently present to consciousness, from which one reaches other knowledge by inference and hypotheses. In the other, used by English empiricists and utilitarians (i.e. pragmatists) from Locke through Hume to Spencer, it means the primitive base which has been transformed for pragmatic reasons in the course of development to give our present consciousness. Bergson, he thinks, adopts the Scottish sense when talking of perception, memory, and freedom in *Matter and Memory*, but the English sense when talking of the nature of intelligence and of how we get the notions of intensity or of number and homogeneous time in *Time and Free Will*. Similarly Čapek, distinguishes what he calls immediacy *de facto* and *de jure*, where the latter is what is left when introspective data are freed from irrelevant and extraneous elements. (Čapek 1971: 86 – 8.) Čapek's *de facto* sense seems to correspond closely to Berthelot's Scottish sense, and the *de jure* and English senses, though introduced in different ways, seem in the end to point to the same concept, though Čapek's way of introducing it seems better: it is not the concept's pragmatic history, if any, that is important.

The possibilities for error if one doesn't make this distinction are clear. It is not at all obvious that there is such a thing as the *de jure* immediate, as the history of the sense datum theory and the search for 'foundations' of knowledge have shown. On the other hand one must start somewhere, and if the *de facto* immediate is simply that which one takes for granted to start with then at least one must assume there is somewhere one can so start, though one cannot assume that everyone else will start from the same place. It is not even necessary that two people in discussion must start from the same place. A might assert something, B contradict it, and A admit he was wrong: the resulting agreement is not a common *starting* place.

It is indeed not clear that Bergson distinguishes the two senses. Answering the question 'why one should accept without reserve as true and real the ultimate data of our consciousness' he replies, 'Because any philosophy whatever must start from these data', and proceeds to take as examples our immediate consciousness of free will and of movement (M, 771). Berthelot compares this to an astronomer saying that the apparent movement of the stars must be real because we must begin from them. (Berthelot 1913: 329 – 30.) This, however, is not quite fair on Bergson, who makes clear elsewhere that the acceptance need only be provisional: 'all natural belief should be held as true, all appearance taken for reality, as long as its illusory character has not been established' (CM, 45 (40)).(Cf. M, 200; both passages were admittedly written long after Berthelot's book.)

What suggests that Bergson is here primarily thinking of the *de jure* sense is the emphasis he puts on the difficulty of achieving intuition. The 1915 letter already quoted talks of the intuition of duration as requiring 'a very great effort' and 'something like a new method of thought (for the immediate is far from being what is easiest to perceive)' (M, 1148). (See M, 796 – 7, for a similar letter in 1909.)

7. Concepts and language: the problem

So far we have looked at two features of Bergson's view of intuition which do not fit all that obviously together. One is the notion of immediacy, and in particular what I have called *de jure* immediacy, following Čapek, and the other is the way in which things can be known by 'sympathy' from the inside, as we experience duration or a melody or a quality; as Bergson says, talking of intelligence, 'no new *quality*, no object of simple intuition, will come from there' (M, 772; italics his). The trouble is that it does not seem to require prodigious effort to experience these things. What might require effort is to come to a philosophical conclusion either concerning or on the basis of such experiences. But experiences are not conclusions. In fact we seem to come back to the notion of knowledge. When we discussed instinct we were concerned only with those instincts that seem to involve knowledge, as opposed to those that make us duck to avoid a blow. The knowledge in question was of the sort that could be expressed in propositions, and it seems to be propositional intuitions that we need. Intuitions after all are to be the basis of philosophy and philosophy is presumably propositional.

Suppose I have long known about pineapples but only now taste one, saying to myself, 'So that's what pineapples taste like': have I acquired propositional knowledge? Simply experiencing the taste itself may be a new experience but is not knowledge. But I brought the experience under the concept pineapple – I knew it *as* the taste of pineapple – and this concept is public and goes beyond the present token experience. I might be wrong in supposing this is what pineapples in general taste like, even if this one does. This suggests I have acquired propositional knowledge.

We seem to be going at least in the direction of a philosophical role for intuition. Unfortunately we are also going away from what Bergson actually says, for we have appealed to two notions which he explicitly associates with intelligence as against intuition, namely concepts and language. I have already quoted Bergson as saying how static knowledge by concepts separates knower from known (M, 773). Concepts are the instrument of intellect because they are 'outside each other, like objects in space; and they have the same stability as such objects, on which they have been modelled' (CE, 169); he goes on to say they are not images but symbols. (Cf. CE, ix.) What is not very clear is how they could be modelled on objects if we require them to pick out objects as such in the first place – for objects after all depend for their reality *as* objects on being picked out by us for pragmatic purposes. Presumably the point is that animals, which certainly need to act and therefore to pick out objects, have concepts of a kind (insects have language of a kind at CE, 166), but not yet the full-blooded concepts which enable humans to have a science of objects. There seems to be a certain chicken-and-egg puzzle here which resembles those concerning the development of language: how could subjects develop without predicates to apply to them or predicates without subjects to apply to? This kind of dilemma is not peculiar to Bergson.

The position concerning language is obviously related to that concerning concepts, though in one respect less simple: it is not always clear whether it is language as such that is incompatible with a correct intuitive view of things, or simply our actual language. Many philosophers have held that language can lead us astray without claiming it inevitably must do so or wishing to transcend it altogether. In a passage we discussed in connexion with change in chapter 4 Bergson says that if language were 'moulded on reality we should not say "The child becomes the man", but "There is becoming from the child to the man"' (CE, 330); and we have just seen that insects, the prime bearers of instinct, can have language.

Elsewhere, however, language itself seems to be at fault. *Creative Evolution* ties perception, intellection and language to the cinematographical approach (CE, 323; see also CE, 344–5). In a parenthetical remark we are told, 'we must adopt the language of the understanding, since only the understanding has a language' (CE, 272). Even more explicit is 'Introduction to metaphysics', where a discussion of intuition near the beginning culminates in the italicised sentence: 'Metaphysics is therefore the science which claims to dispense with symbols' (CM, 191 (162)). ('Science' is being used in the early and wider sense it has in 'Introduction to metaphysics'. See also CM, 229–30 (194–5).) This approach is also illustrated in William James's interpretation of Bergson on concepts, which Bergson greeted as 'perfectly exact' (M, 821). 'Nothing *happens* in the realm of *concepts*', says James (1909: 340), and only perception, not conception, can give us insight into moving life. Bergson 'limits conceptual knowledge to arrangement, [which is] the mere skirt and skin of the whole of what we ought to know' (James 1909: 343). James gives an excellently clear and readable summary of material scattered in Bergson, but does not, I think, resolve the sort of difficulties we are discussing, though he does answer some objections.

8.Immediacy and the role of intuition

How can metaphysics or any other science, even in the widest sense, dispense with language? Well, perhaps we abandoned prematurely the search for *de jure* immediacy. Our argument has been this: The experiencing of qualia does not seem to require effort. That is only needed when we go on to form propositions concerning or on the basis of such experiencing and the search for propositions involves us in language and symbols, and metaphysics is supposed to dispense with these. But is it true that the experiencing of qualia never involves effort or training? How about the experience of hearing the inner parts of a string quartet? And how about the training that a painter needs if he is really to 'paint just what he sees'? Even if ultimately there is no such thing as 'just what he sees' there is no doubt that painters can be trained to see both a great deal more and a great deal less than the rest of us can see.

Though this does not get over the objection about philosophy using language, it may make some progress. In the same 1915 letter quoted above Bergson himself compares his method to artistic intuition – or rather deprecates the comparison, on two grounds: his method concerns itself with matter as well as life, and so appeals to intelligence as well as intuition; and though philosophical intuition goes in the same direction as artistic intuition it goes much further: 'it takes the vital before its scattering into images, while art bears on the images' (M, 1148). The obscure second point, which he does not elaborate, suggests that philosophy somehow grasps life as it is in itself rather than giving a representation of it in the way that an artist might. This in turn suggests at least some affinity with the view that metaphysics dispenses with symbols.

A.D. Lindsay (1911: 226ff.) compares Bergson's doctrine of intuition with that of Plato as exemplified in his *Seventh Letter*, where he says of his own metaphysic that it 'cannot be put into words as can other enquiries' (quoted at 239) but suddenly springs out after long study, which suggests to Lindsay that 'Plato is one with Bergson in insisting that true knowledge must dispense with symbols' (227). He quotes a long passage from Bergson comparing metaphysical intuition to the sort of impulse one gets to literary composition after strenuously soaking oneself in the relevant materials (Lindsay 1911: 237–8). (Cf. CM, 235–6 (199); Lindsay uses a different translation.) Unlike Plato Bergson says 'there is nothing mysterious about this faculty' (ibid.), but no doubt there is something common between the two. Intuition is seen as the getting of bright ideas, be it in metaphysics, science, mathematics, literary composition, or anywhere else, and it is notoriously impossible to provide any algorithm for achieving such ideas. One might compare the way Plato in the *Republic* sees such importance in the notion of dialectic while saying so little about it.

In discussing the place of mathematics in Bergson's system Lindsay rightly distinguishes the use of a tool from its invention, which requires 'the insight of genius' (Lindsay 1911: 224). He adds that 'when Bergson says that intelligence uses concepts like tools he does not mean that the concepts themselves are the work of intelligence as he describes it' (ibid.). They require 'some vivifying insight' (222). Well, maybe. But in so far as concepts are tools we shall expect intelligence to make as well as use them: 'intelligence . . . is the faculty of manufacturing artificial objects, especially tools to make tools, and of indefinitely varying the manufacture' (CE,

146); it is the faculty of 'making and using' unorganised instruments (CE, 147). (Cf. CE, 158.) At p.220 Lindsay agrees, but still goes on to link intelligence with the mechanical and what can be reduced to rule. This is no doubt correct; but how consistent a position does Bergson have? He attributes the invention of the infinitesimal calculus to intuition (CM, 225 (190)) and then says that 'it is the symbol alone which intervenes in the application. But metaphysics, which does not aim at any application, can and for the most part ought to abstain from converting intuition into symbol' (CM, 225 (191)). (Lindsay, quoting from another translation on his p.225, omits 'for the most part'.) Now as Lindsay (1911: 224) points out mathematics needs imagination and insight and is far from being mechanical. But it needs these in applying concepts almost as much as in inventing them, as the calculus well illustrates, since there is no general algorithm for integration. In fact any science or any study worthy of the name of intelligence will require this sort of insight beyond the most elementary level, and it is this which gives intelligence the flexibility that distinguishes it from instinct. *This* could hardly distinguish metaphysics from other studies. It is true, as Lindsay well says (1911: 218, 241), that Bergson tends to isolate and push to extremes tendencies he admits could not really occur apart from each other. But the sort of insight we are discussing is hardly an inevitable but alien intrusion into the sciences Bergson assigns to intelligence; it is part of their very essence. The most that might be argued is that the sort of algorithms and deductive techniques that are common tools in the other sciences are totally absent from metaphysics, something perhaps truer of the metaphysics of Bergson's day than of the more recent analytical orientation. But even this idea could only be taken a short distance. Even the airiest of metaphysics can hardly subsist entirely on the getting of bright ideas without drawing any implications from them or tracing any connexions between them; certainly Bergson would not want his own philosophy to be of such a nature: 'my intuition is reflection' (CM, 103 (88)).

The dilemma we have reached is this: on the one hand intuition is the tool of metaphysics, which dispenses with concepts and symbols and has life for its field of study. Intellect is the tool of science, which uses concepts and symbols and has matter for its field of study. On the other hand when we examine intuition it seems to be, or at least to include as one of its forms, something that is as necessary for science as for any other study, while it is absurd to speak of metaphysics itself as dispensing with concepts and language; and, just as science cannot do without insight, so metaphysics cannot do without the intellectual activity of argument. A further problem arises: science and metaphysics are each competent in their own field; they each give us 'absolute' knowledge provided they don't trespass on each other's fields, and Bergson vigorously insists that he is not decrying science or subordinating it to metaphysics (e.g. M, 747). On the other hand 'absolute' knowledge knows its subject 'from within', but to do this is the province of intuition, so how can science have absolute knowledge of matter? Here we touch again on the question how the intrinsic distinction between the faculties is related to the distinction between their objects. There is yet a third problem: if life is the subject of metaphysics, what happens to the biological sciences?

To start with the third and easiest problem: the note at M, 747, makes clear that the biological sciences are 'relative', because they approach problems of life and consciousness from the point of view of physics and chemistry. They are 'legitimate' but need to be supplemented by metaphysics. The other secondary problem has something to do with the change in the use of 'intuition', which originally (in 'Introduction to metaphysics') covered absolute knowledge in general and later (in *Creative Evolution*) was restricted to knowledge of life, though Bergson says we must look to the context in interpreting it (CM, 306 note 26 (Q, 1423–4)). Perhaps the point is that the only way in which matter is susceptible of being known from within is the way in which science would know it, with its concepts and measurement. This would not be surprising, since life might be expected to be a more suitable subject for being known from within by intuition, whose possessor is himself a living being; no doubt he is also a material object, but he is not exercising his materiality in doing his intuiting.

9. *Concepts and language: development of the problem*

We can now return to the main dilemma. What really is Bergson's attitude towards concepts and language? In his 1901 discussion of psychophysical parallelism he describes language as the instrument *par excellence* which liberates man from the level of mere animality, 'despite the automatism which it inflicts on thought in the long run' (M, 487). Here again it seems to be language as such that is defective. But two pages earlier, where he describes his proposed philosophical method as that of 'following the real in all its sinuosities', he adds that it does not consist in 'extracting from reality a simple concept (or rather a negative concept like that of non-parallelism) to be then submitted to being worked on by dialectic' (M, 485). Our intelligence favours proceeding by 'stable conceptions' (CM, 222–3 (188–9)). It is tempting to take these adjectives, 'simple' and 'stable', as limitative, suggesting that it is not concepts as such but only concepts treated as frozen and unchanging that are hostile to intuition; one might add passages where to *analyse* duration is to resolve it into 'ready-made' concepts (CM, 217 (184)), or where again it is 'ready-made' concepts that give the trouble (CE, 51). But there is little really to support this. The main passage is perhaps one where intuition 'will arrive at fluid concepts, capable of following reality in all its windings and of adopting the very movement of the inner life of things' (CM, 224 (190)). (J. Milet uses this; see §11 below.) But this isolated passage seems inadequate to oppose the large number of passages treating concepts as foreign to intuition. We might think it represents what Bergson *ought* to have said, opposing such fluid concepts to the 'certain classifications which have been made, once for all, with a view to our action upon matter' which in fact in 1911 he equated with concepts as such (M, 946). His division of concepts into natural and artificial does not affect this issue; in a lecture at Columbia in 1913 he is reported to have first made this distinction and then gone on to examine the possibility of a 'consciousness without concepts, that is to say intuitive' (M, 977).

The truth is that Bergson does not see the relation between concepts and thought as we might see it. At least for most of the time, with the possible exception of the 'fluid concepts' passage, it is concepts as such that he is talking about, not some subset or special kind of concepts that are more rigid than the rest; but he does not think they are essential for all thought. Not only can intuition dispense with them, but in an important note in 1910 where he tries to clear up some misunderstandings he says that even for science itself concepts are more 'provisional schemas'; science really aims at resolving matter into mathematical relations (M, 824). Science, the same note tells us, does not deal with the unreal, even though objects are unreal in the sense of being artificial constructions; this includes scientific objects like atoms and fields of force, though probably not the organic. (See CE, 240.) It is qualities and change that are real, and as different creatures may perceive these at different degrees of tension, as we discussed in chapter 5, science studies them at their lowest degree of tension, where what perception sees as a red light becomes reduced to trillions of vibrations.

10. The nature of philosophy

We are still left with the problem of how science, metaphysics, or anything else that uses language can go beyond concepts. Do not concepts and language go hand in hand, and does Bergson really think we could do metaphysics without language? The answer to this last question is not very clear. Obviously Bergson uses language himself in his metaphysics, but philosophers often put forward philosophies inconsistent with their being put forward. 'Whereof one cannot speak thereof one must keep silent,' said Wittgenstein, at the end of his *Tractatus*, but only after he had done his share of speaking. However, some further light is thrown on Bergson's position when he asks whether his philosophy involves a circle in using intelligence to transcend intelligence (CE, 202–10). If this really were an objection, he replies, it would prove too much. We could never learn to swim, it might be thought, because we must first know how to hold ourselves up in the water. But we do learn to swim – by swimming. Similarly we can get from intelligence to intuition – by plunging in; as he often says, we must place ourselves in the midst of duration at the outset. What we cannot do is get to intuition by intellectual means, because the reality to be intuited always contains more than is accessible to intellect. As we saw when discussing Zeno, intellect can divide the trajectory as finely as you please, but can never for Bergson give us that extra element, the transition from one instant to another. We can, however, move in the opposite direction. Once we have experienced the motion, intellect can set to work on the trajectory. (CM, 213 (180). Cf. the summary of his 1901–2 lecture course at M, 517, and CM, 215–16 (183), 223–4 (189).)

What is left in some obscurity is what this plunging in is supposed to consist in, and what exactly is supposed to result from it. Bergson goes some way towards telling us at CE, 204–10, where he discusses the place of the biological sciences, which as we saw were 'relative'. Physics belongs unambiguously to intelligence and metaphysics to intuition, but how about the science (which suggests intelligence) of life (which is the object of intuition)? The answer is that so long as its aim is practical, to enable us to manipulate living things, it is simply an extension of physical science, but if we want theoretical knowledge of life only philosophy can give it to us. This involves philosophy itself in entering a new domain: 'She busies herself with many things which hitherto have not concerned her' (CE, 209). At the same time philosophy becomes empirical, at least in part: 'philosophical method, as I understand it, is rigorously founded on experience (internal and external)', as a 1912 letter put it (M, 964). (Cf. ME, 59, and also M, 479.)

Philosophy can give us concrete explanations of particular facts as well as general theories, and can be as precise as positive science, he wrote in 1915, with reference to *Creative Evolution* (M, 1181n.). All this ecumenical writing is all very well, but though philosophy should 'superpose on scientific truth a knowledge of another kind, which may be called metaphysical' (CE, 209–10), we still haven't been told much about what this knowledge really consists in. It precedes science, as the 'bright idea' which set science going, and follows it in the sense that it requires a lot of scientific preliminaries, again a feature of the 'bright idea'. (CM, 235–6 (199), 226 (192); we need not on this account see an ambiguity in it with Čapek (1971: 88).) But this raises a certain problem, for soon after the passage we quoted earlier about how metaphysics ought 'for the most part' not to convert intuition into symbol we are told that 'intuition once grasped must find a mode of expression and application which conforms to our habits of thought and which furnishes us, in well-defined concepts, the solid basis . . . we so greatly need' (CM, 226 (191–2)); similarly the summary of the 1901-2 lecture mentioned above ends by telling us that to avoid arbitrariness we must not lose all contact with conceptual thought, and our results must always, as far as possible, be translatable into concepts (M, 517). So we can express our metaphysics in language after all? It is not clear. The context in *The Creative Mind* suggests it is the science following after the 'bright idea' that is being referred to, and the remark that thought, whether intellection or intuition, uses language may mean the same (CM, 39 (35)). On the other hand:

> Dialectic is necessary to put intuition to the proof, necessary also in order that intuition should break itself up into concepts and so be propagated to other men; but all it does, often enough, is to develop the result of that intuition which transcends it. (CE, 251)

Perhaps here we do see a real ambiguity in intuition, for bright ideas indeed can be confirmed by intelligence, even if unaided (algorithmic) intelligence could never have reached them. But how could intelligence possibly confirm an insight of a kind fundamentally inaccessible to it? And let us consider again the case of mathematics: we saw above that the use as well as the invention of the calculus involves intuition in the 'bright ideas' sense. But it is intuition in the other sense, intuition of duration, which is supposed to solve Zeno's problem – which it doesn't: the mathematical theory of the infinite, using concepts like those of *open set, limit,* etc., is an intellectual construction owing nothing to ineffable intuitions. Even if it involves 'abbreviations' of statements that could never be written out in full, there is no room for intuitions in any sense beyond that in which they may be needed for any use of intelligence at all. (See here, and also for some of the criticisms mentioned below, Cunningham 1916: chapter 3.)

Here then is an ambiguity in Bergson's view of intuition. Sometimes it is an ineffable means of acquiring perfectly effable knowledge. At other times it is a source of 'knowledge' ineffable in its nature. But there seems to be another ambiguity too. I have called this chapter 'Epistemology' but Bergson's first work devoted primarily to its topics he calls 'Introduction to metaphysics'. For us epistemology and metaphysics are two separate if related branches of philosophy. But Bergson nowhere, I think, divides philosophy like this. Intuition is a mode of acquiring knowledge for Bergson and philosophy is *par excellence* the subject which uses it. But what is the subject which studies it, and what does that subject use to do so? Most of the time he writes as though the job of philosophy was to use intuition, in the second sense above, and tell us the results of doing so, which leads to the difficulty that this telling can only be done in language, which can only freeze the results into concepts and so distort them out of their proper nature and give us something other than what we were supposed to be given. But Bergson is a philosopher and uses language to write philosophical books. What these books tell us is *about* intuition. They do not give us intuitive knowledge but tell us something of what intuitive knowledge is and how we might expect to reach it. They do not give us the experiencing of red or duration but tell us *that* the experiencing of these is different from anything that science could tell us. They tell us to enter the Promised Land but do not lead us into it. This is true even of the 'metaphysical' books like *Matter and Memory* and the non-methodological parts of *Creative Evolution*, as well as 'Introduction to metaphysics', which is really an introduction to epistemology. But, since no book could lead us there, what reason have we to think that there is a Promised Land? For qualia and duration are things we intuit already, and we could hardly be alive at all if we didn't, while the immediate in the *de jure* sense is a doubtful thing at best. What often seems to be hovering in the background is some sort of mystical experience, and it is not surprising that in his last main book, *The Two Sources of Morality and Religion,* Bergson gets much nearer to a mystical view. (See MR, 219–20.)

11. Conclusion: the role of mathematics

Let us draw things together and discuss one final point. It is pretty well *de rigueur* by now for anyone writing on intuition and intelligence in Bergson to point out that he was not an anti-intellectualist. He repudiated the label himself (M, 756) – though he could also say ' . . . je sympathise avec l'anti-intellectualisme de Sorel' (Mossé-Bastide 1955: 96). L. Husson sees him as an intellectualist (Husson 1947). J. Milet goes further, following C. Péguy, and calls him a rationalist in two senses (Milet 1974: 100ff.) (Cf. the short English résumé in Milet 1987, but also the review in Theau 1976.) How far should we agree?

We have seen two senses of 'intuition'. In one it means seeing things from within. In a weak form, presumably shared with animals, we exercise it in seeing trillions of vibrations as the quale red: 'quantity is always nascent quality' (CM, 225 (191)). But more properly it involves the appreciation of duration; animals probably don't appreciate melodies quite as we do. In this sense it develops into the method of a particular study, metaphysics, and has a particular object, life or spirit. In the other sense intuition is something like the getting of bright ideas, essential for any intellectual work and useless unless followed up by intellectual labour, which it also presupposes. Emphasising the first sense may tempt us to call Bergson an anti-intellectualist; emphasising the second may tempt us the other way. Can they be reconciled?

Milet argues that Bergson was influenced by mathematics from his earliest days, and during his early years at Angers and Clermont (1881–8) came to see in the calculus, especially as developed by Newton in terms of 'fluxions', a method of getting to grips with duration and movement. Milet relies primarily on CM, 224–7 (190–2), with CM, 37 (33–4), M, 753–8, MM, 241–2, CE, 33–4, and other passages. He argues that mathematics shows the way for metaphysics, which should operate 'qualitative differentiations and integrations' (CM, 226 (191); translation altered). Elsewhere 'vitality' relates to physicochemical forces rather as curves to their tangents, which however short cannot compose them (CE, 33). Bergson's new rationalism relates to classical rationalism as the infinitesimal calculus to the classical calculus (i.e. pre-calculus mathematics); as number and shape are particular cases of a dynamic process which engulfs them, so ideas, judgements, and reasonings are particular cases of dynamic mental processes which engulf and integrate them: this is the nearest Milet comes to explaining this analogy between mathematics and metaphysics (Milet 1974: 102–3). He develops an interpretation of Bergsonism in some detail, summarising it at p.126, but his claim to reveal it as 'a rationalism, in the most rigorous sense of the term' (106) seems optimistic.

What never seems to be resolved is the clash between the senses of 'intuition'. Modern mathematics tries to replace the ready-made by what is becoming, yet even there it falls short of metaphysics because it 'contents itself with the pattern, being but the science of magnitudes' (CM, 225 (190)). Similarly for *Time and Free Will* the calculus concerns only the extremity of the interval never the interval itself (TF, 119–20). (Both passages are referred to by Theau (1976)). Modern mathematics gets close to its object in a way that biology probably never will (CE, 34). Yet intuition seems to come in only with the invention of the calculus. This alone is metaphysical. But as soon as we turn to its application symbols come in and mathematics becomes the science of magnitudes, while metaphysics needs no application and therefore ('for the most part') no symbols; intuition is essential for science, but is a generative act which 'lasts only an instant', and rather than being peculiar to metaphysics is where science and metaphysics meet (CM, 225–6 (190–2)).

Evidently in all this the 'bright ideas' sense of 'intuition' is dominant, though the link with metaphysics is there. That Bergson began from mathematics and mechanics rather than psychology is indubitable. (See the autobiographical note at Q, 1541–3.) But a letter of 1903 makes clear that thinking about duration led him *from* the former *to* the latter (M, 604). Milet quotes this, but it may not confirm his view as fully as he thinks (Milet 1974: 34–5). A conversation in 1911 suggests Bergson had thought mathematics could not deal with duration, and so turned to psychology instead, not as a field to apply the lessons of the new mathematics (Chevalier 1959: 19–20; cf. 228). Admittedly Chevalier, who reports the conversation, is a far lesser thinker than Bergson, and may distort him on some issues; but he has no motive to do so in the present case, and Bergson, a life-long friend, approved Chevalier as an interpreter of him in general.

Milet thinks that Bergson sees mathematics only as a model for metaphysics, not as replacing it. Mathematics remains a science of quantity, but its 'manner of thought' is 'thinking in duration', which metaphysics should adopt. (See especially Milet 1987: 34.) I will make three comments.

First, this is surely not what led Bergson from mathematics to psychology, but exists as an independent undercurrent. As elsewhere (e.g. on whether duration belongs in the external world) Bergson vacillates. Alongside Milet's evidence we have the bald statement that 'duration, as duration, and motion, as motion, elude the grasp of mathematics' (TF, 234), to say nothing of TF, 119–20, mentioned three paragraphs above, and the later CE, 23, in the context of an explicit discussion of the calculus: 'the world the mathematician deals with is a world that dies and is reborn at every instant'. See also Robinet 1965: 20–1, 148–9. It is hard on any view to see how mathematics can 'seize motion . . . from within' and 'adopt the mobile continuity of the pattern of things' while it 'contents itself with the pattern' (CM, 225 (190)).

Second, for Bergson it was astronomy rather than mathematics that introduced time into science: 'Modern science . . . has come down from heaven to earth along the inclined plane of Galileo' (CE, 354). However, science as such treats of time rather than duration, yet for centuries science has used the calculus, so in what sense does the calculus really contribute to our grasp of duration? (See CE, 355, 364.)

Third, 'the essence of Cartesian geometry . . . was to regard every plane curve as described by the movement of a point' (CE, 353). This only gives us time, and the calculus is needed for us to get duration, if anything will give it us. (Cf. CE, 361.). But it points to a possible error on Bergson's part. The calculus helps us to deal with continua, some but not all of which involve time. But it involves approximation to a limit, and it is tempting to think that because we, if we approach a limit, must take time to do so, therefore the calculus itself somehow essentially involves time. Milet represents Bergson as thinking of 'infinitesimal thought' as apprehending the continuous not 'all at once' but progressively, and as deriving 'thinking in duration' from this (Milet 1974: 146). If Milet is right in his interpretation, surely he and Bergson alike are wrong philosophically. (Nor does the introduction of infinitesimals by modern 'non-standard analysis', due to Abraham Robinson, seem to make any difference here.)

Finally, returning to intuition, which Bergson admits is ambiguous (CM, 37–8 (34)), perhaps one thing Bergson thinks is that any getting of bright ideas, in any subject, is *pro tanto* metaphysical, the relevant science developing the idea in symbols. But, be that as it may, Bergson hardly succeeded in explaining what 'qualitative differentiations and integrations' would amount to (CM, 226 (191)). Indeed he hardly tried.

VII

Biology

1. Introduction

Creative Evolution, as its title suggests, is a book on biology, despite being much more. Just as Bergson was to grapple with Einstein on physics so here, with rather more confidence, he enters the lists with the scientists of his day on the topic of evolution, which had come to prominence so dramatically in the year of his birth. The main discussion of evolutionism as such, or transformism as he calls it, comes in the first chapter of *Creative Evolution*, where he discusses the various alternatives, Darwinian and other, and introduces his own theory. The rest of the book develops this by outlining the course evolution has actually taken, especially as it affects the emergence of instinct, intellect, and intuition, and then goes on to elaborate the resulting philosophical themes some of which we have discussed in chapter 6. We have in fact reversed Bergson's own procedure by looking first at his general epistemological and methodological position before seeing what happens when this is applied to science itself; the alternative is to treat the philosophical position as resulting from the scientific one, in the sense that the distinctions between instinct, intelligence, and intuition, or rather the manifestations of these distinctions in the actual world, are what evolution in fact leads to. The former course seems preferable, since his views on evolution, though given some empirical support, plainly arise from his general philosophical position, and it is as his views rather than as the truth about evolution that we are primarily approaching them. For the same reason we need not feel as constrained as he felt himself to be by his otherwise excellent advice to unearth the empirical facts as science now sees them; this is not the place for an enquiry primarily devoted to the actual facts about evolution, especially those discovered after he wrote.

A sort of motto for *Creative Evolution* occurs in a remark Bergson made in 1908 when discussing ways in which knowledge can be qualified. After insisting that knowledge can be limited to a certain part of its object without ceasing to be 'absolute', in the sense we discussed in chapter 6, he goes on: 'Now, one of the objects of *Creative Evolution* is to show that the Whole is . . . of the same nature as the self, and that one grasps it by an ever more complete understanding ['approfondissement'] of oneself' (M, 774).

In other words the world is not something set over against the self and of a different nature; it shares the nature of the self and is itself in some sense spiritual. Not surprisingly it will fall within the sphere of intuition to study it, but the difficulties that leads to we have already discussed.

2. The nature of life

In view of the then still recent, if fading, opposition to evolutionism on religious grounds it may seem significant that so 'spiritual' a philosopher as Bergson should accept it so wholeheartedly. But it does of course fit in very well to a philosophy of change, and especially to one involving the notion of duration with the asymmetry that this involves. For the inorganic 'the present contains nothing more than the past, and what is found in the effect was already in the cause' (CE, 15); when life comes on the scene this no longer holds, and evolution is a one-way process. Actually the quotation is applied to 'unorganised bodies', but this must include the inorganic, since his example is a chest of drawers. A later passage suggests that 'unorganised' does not mean 'unstructured', as a stone might be but a chest of drawers would not, but contrasts with possessing organs as a living body does (CE, 36). Bergson's real contrast is between living and non-living. The organic but non-living forms a sort of penumbra round the living, but is presumably parasitic on it, so that we need not always distinguish them – a later summary contrasts 'inorganic' and 'organised', apparently as equivalent to 'inert' and 'living' (CE, 196).

However, a problem arises. We saw in chapter 2 how Bergson started by confining duration to consciousness, but then (from *Matter and Memory* onwards) allowed it to the world itself, for consciousness must live in the world, and how can that which has duration interact with that which has not? But the same question arises about life, if irreversibility or duration is a mark of this while time has no significance for the non-living, which could be speeded up indefinitely without affecting its nature. Again how can one part of a single world have duration and another part not? Bergson's answer seems to be that it is only when considered in isolation, as it is when studied by mathematics or physics, that the non-living is indifferent to time (CE, 10 – 12). As soon as we take it in its context of the world as a whole, which contains life as well, and interacts with consciousness, and endures, the non-living too endures. His example is that of sugar melting in a glass of water: we have to wait while it does so. (See chapter 2 §2 above and also CE, 358; salt would be an inorganic example.) Presumably the point is this: first, we might accelerate the melting by adding heat, but this is irrelevant; we would still have some period to wait. However, we could *imagine* it speeded up without interference, provided that everything around it (e.g. the nearby clock) was speeded up equally. But if we add ourselves into the imagined situation, who are in duration intrinsically and so cannot even in principle be speeded up without alteration, then neither can the sugar.

We discussed the speeding up argument earlier, but in the present context, despite the example, the emphasis is rather on irreversibility than on the absoluteness of the rate of change. Organic, and in particular evolutionary, changes are irreversible while ordinary movements of matter are not. This criterion for distinguishing the living has been criticised, along with two other criteria Bergson uses, by Berthelot, who points out that things like mountains and stuffs like steel go through irreversible changes (Berthelot 1913: chapter 12, especially 283 – 6). A mountain has a definite evolution, being, say, spiky in its youth and rounded in its old age. The Cairngorms probably once looked like the mountains of Iceland, which they certainly don't now. Steel, or paper for that matter, which is folded and then unfolded does not return to its former state, but carries a 'memory' of what has happened to it.

The point is a good one, and Bergson does not discuss it. How might he answer it? He might suggest it is merely contingent that mountains go through the cycles they do; a suitable combination of climatic and geological circumstances could return the Cairngorms to their former spiky state, even if it never will. Could the sun return to a state where less of its hydrogen had turned into helium, or the ripples on a pond converge back upon the stone that caused them, and the stone itself rise out of the water? Presumably yes, if a sufficient energy supply were forthcoming, and presumably too the same would hold of the creased paper.

What is true is that the energy supply might have to be very differently distributed from that which caused the original phenomenon, and the required distribution would be a very improbable one. Suppose we found a strip of film depicting the movements of a gas in a confined space. At one end (call it E) is a frame showing the molecules bunched in one part of the space. The frame at the other end (call it L) depicts them roughly evenly distributed. The intermediate frames depict intermediate distributions. The differential equations of physics by themselves cannot tell us which way to run the film, since a differential coefficient represents the tangent to a curve at a point and such a tangent goes in two directions from the point. It is the law of increasing entropy (the second law of thermodynamics, or Carnot's law, as Bergson calls it) that tells us which way to run the film, and this law is statistical. Roughly it says that, for any state of the gas depicted by a given frame F, if we consider all possible states the gas might be in shortly after this state a far greater proportion of them will resemble state L, i.e. will show the gas more evenly distributed, than the proportion which will resemble state E. In this sense it is easier for the gas to pass from state E to state L than vice versa, and so we call E early and L late.

So the inorganic changes can be conceived as reversed if we forget about entropy, and the position is the same for the organic. A chicken could retreat into its egg except for the entropy law, for the chicken, like the gas cloud, is but a set of molecules as far as physics is concerned.

That a chicken *is* just a set of molecules is what Bergson would hotly deny. But he cannot give facts about irreversibility as a reason for denying it, even if he could infer such facts after denying it on other grounds. Berthelot similarly argues that two other criteria Bergson uses to mark out life (or animal life in the case of the second: CE, 126), adaptability and the ability to store energy and suddenly release it, are not confined to the organic (Berthelot 1913: chapter 12).

Perhaps then Bergson would have done better to draw his main distinction between consciousness and its absence rather than between the organic and the inorganic, or the living and the non-living, where apart from anything else he is always vulnerable to advances in science. This would indeed fit in well with much of the spirit of Bergson's philosophy, but not with all of it. We have already seen something of his approach to consciousness and of the extent to which he was and was not a dualist. Consciousness for him was not, or should not have been, something which subsisted on its own, however 'spiritual' his philosophy may have been. This is mainly because of its admittedly ambiguous relation to choice and action. Action requires something to act on and hence a material world. It also requires something to do the acting and hence bodies of some sort. But living bodies have properties that go beyond consciousness. Not only may they digest food and circulate blood but many of them are vegetables, which share with animals the two properties that marked the minimum grade of life for Aristotle, the ability to preserve and within limits develop one's form through metabolism and the ability to reproduce that form. This last property is obviously essential to any candidate for being the subject matter of evolution. It is true that Bergson officially regards consciousness as coextensive with life (CE, 196), but only by plunging the vegetable kingdom into almost complete sleep: 'movement and consciousness sleep in [the plant] as recollections which may waken' (CE, 125). (Cf. CE, 111 – 26.) This suggests they were awake previously, and there is a distinct awkwardness, if no worse, in Bergson's thought here. Animals presuppose plants for nourishment, while plants nourish themselves directly from minerals, so surely plants must come first, in which case how can they or their ancestors have had the experiences that could give rise to 'recollections'? His answer in effect (he does not give it as an answer to this particular problem) is to say that plants and animals diverge from a common stock, which 'oscillated between the vegetable and animal form, participating in both at once' (CE, 118). Presumably these primitive organisms fed partly on minerals and partly on each other; but as they were presumably simple as organisms go (for how could complex ones precede simple ones?) the mobility and accompanying consciousness that they had *qua* feeding on each other could hardly be more than rudimentary, and would be a slender grounding for whatever it is that 'sleeps' in vegetables generally.

Bergson is always aware that sharp distinctions or rigid definitions of terms like 'life' or 'plant' and 'animal' are difficult if not impossible, and it is realism rather than laziness that makes him insist on the fuzziness of all such boundaries. Plants are unconscious – but not always: the lowest plant forms, paradoxically, such as the zoospores of the algae, have rudiments of consciousness to the extent that they have *liberty* of movement (CE, 118, my italics). But the only evidence he gives of such liberty is that they 'hesitate between the vegetable form and animality' (ibid.). Is not the 'hesitation' Bergson's in classifying them? His flexibility in developing his ideas and his willingness to accept implausible conclusions if his argument requires them are praiseworthy enough, but it is hard to deny that the role he gives to consciousness as the dominant element in the sphere of life is forced and is given in the interests of the kind of vitalism he is going to develop as his own theory of evolution.

Obviously the task of defining life is a perilous one. Two paragraphs ago I mentioned the property of self-reproduction as essential for the subject matter of evolution to possess. One might be tempted to define that subject matter in terms of it, but that would be unfortunate if life is what we intend to capture, since structures made of bits of wood, like a jigsaw, can be made to reproduce further structures of the same shape if shaken in a box containing a large number of loose pieces, provided they have the right shape. (See Moore 1964.) Whether or not we want to count the first molecules to reproduce themselves on the primeval earth as living, we presumably do not want to count bits of wood or metal etc. as doing so just because they have a special shape.

In chapter 4 I said that for Bergson 'the division of the world into objects is an artificial division made by agents for practical purposes' (p.96), and went on to discuss how innocuously this was to be interpreted. We now find that living bodies are granted individuality in a sense in which the non-living lacks it (CE, 12 – 24). Like the other terms we have been discussing this individuality is not absolute, for, if it were, 'no detached part of the organism could live separately. But then reproduction would be impossible' (CE, 14). Similarly a creature may split, each half regenerating itself into a separate creature: 'because there are several individuals now, it does not follow that there was not a single individual just before it' (ibid.). What this individuality amounts to is not very clear. It applies to systems that are naturally, and not merely artificially by us, isolated or closed (CE, 15 – 16). (Cf. CE, 318.) But as the discussion goes on the familiar emphasis on duration takes the stage. The point is perhaps this: a living thing contains its past within its present, in the sense that it is essential to it that it has the history that it has; it and its history form a *single* thing which endures, i.e. has duration, whereas 'the world the mathematician deals with is a world that dies and is reborn at every instant' (CE, 23), because the present is affected only by the *immediate* past, but since temporal instants form a dense set there is no instant which is both past and adjacent to the present; hence time in this inorganic context dissolves into a dense set of separate moments, in the way we explored in chapter 2.

Does all this provide us with a mark of the living as such? Perhaps it shows that 'living creature' is a count noun in a way that 'red thing' is not: we cannot unambiguously count the red things in a room because each half of a red thing is itself a red thing, at least until we get to the microscopic level. But this applies to almost anything with an essential structure, given a few exceptions like the papal crown. Bergson claims that, if his chest of drawers contains many pieces after it has fallen apart, it did so before (CE, 14 – 15). So a chest of drawers cannot become many unless it was many already, whereas a hydra, we are told, can become many hydras by fission. But it is not clear whether Bergson thinks this is a fact about the logic of the terms or simply one about the course of nature. The hydra becomes many hydras, which it was not before, but only on pain of ceasing to be the hydra it was. But similarly a chest of drawers could become many chests of drawers provided it ceased being the chest of drawers it was. Even taken as a fact about the course of nature the fact is not very clear: the change in the chest of drawers would normally need an external living agent like a carpenter, though in principle a hurricane could do the job; but even the hydra needs an external food supply. No doubt there are differences between the hydra and the chest of drawers. The hydra, even though it must have food, does seem to have an internal principle of movement that is different from the mere molecular motion that goes on in a chest of drawers. But the differences don't seem to have much to do with individuality. The hydra, treated as a lump of matter, can be dealt with by the mathematician or physicist just as the chest of drawers can, and stars and mountains, however contingently, can go through irreversible sequences; in this sense they too can have a present which was not 'already there' in their past – unless we insist it was 'potentially' there, which would hold for the living too. To avoid confusion, though, we must remember that though Bergson wants to distinguish the living from the non-living this does not violate the 'motto' I attributed to him about the self and the Whole having the same nature. There he is talking of the world as a whole, including both living and non-living, not of the non-living as such.

3.Philosophical arguments

Bergson's arguments for accepting evolution in some form are brief (CE, 24 – 6). He refers to the correspondence between ontogeny and phylogeny, and to the evidence of paleontology, and seems to claim that a classification based on similarity corresponds to one based on phylogeny: if he does mean this, and not simply that animals tend to resemble their *immediate* ancestors (parents), he ignores the differences between what are now known as phylogenetic and phenetic classifications. (See Ridley 1985: chapter 6, and Janvier 1984.) He then adds (CE, 26 – 7) that even if evolution never actually occurred species still bear the same relationships to each other, and still arose in the same chronological order (because of fossil evidence; he doesn't consider Philip Gosse's idea that God put the fossils there to test us). The difference would be that God or some other force created the species separately, though in the same order, and he claims that the philosophical implications would be unchanged. The driving force therefore could have the same effect without working through physiological generation. (See also CE, 56.) The implications of this argument are interesting in view of the way certain religious writers reject not only evolution but also the possibility of things like *in vitro* fertilisation. One wonders why, if God's intervention is needed at all, it cannot occur as easily in the test-tube as in the womb, just as, if God created each species separately, He could have done so by producing it out of a previous species (as Eve came from Adam) rather than out of thin air.

So Bergson accepts that evolution has occurred. But how? Here he adopts his usual policy of contrasting two opposing theories and claiming that they share a common postulate which he rejects, though his own view may be nearer to one side than to the other. The two theories here are Darwinian natural selection, which he regards as mechanism, and finalism, and his own view will incline more to the latter.

The rest of chapter 1 falls into three not sharply separated parts, first, a rejection of mechanism and finalism based on Bergson's own philosophy (CE, 26 – 56), then a rejection of them and one or two other theories on empirical grounds (CE, 56 – 89), and finally the development of his own version of vitalism (CE, 89 – 102).

Bergson insists that biology cannot be reduced to physics, basically because of the time factor, i.e. duration. Science, i.e. physics, the study proper to intellect, requires repeatability and reversibility (CE, 31); and though biological processes can be approximated in the laboratory they remain an ideal limit which can never quite be reached. As an evolutionist he will presumably let the living arise from the non-living, but not mechanically. Laboratories may synthesise things like urea, but these are mere 'waste products of vital activity', not 'the peculiarly active plastic substances' (CE, 36) – though he surely exaggerates in saying 'science has not yet advanced a step towards the chemical synthesis of a living substance' (CE, 38).

The main argument against mechanism is not easy, but depends in the end on the distinction so important to Bergson between a movement and a trajectory (CE, 31 – 41). He also uses the image of a curve and its approximating polygons: 'vitality' is 'tangent, at any and every point, to physical and chemical forces' (CE, 33). Physics and chemistry can only give partial views of life, which will no more reconstruct the whole than a multitude of photographs can reconstruct an object. It is not always clear when we are talking of reconstructing life in the sense of imitating it in a model and when of actually producing it, but we seem to be told partly that we can do neither and partly that we can imitate but that this is not the same as producing; but since the imitation turns out to be physical, not just mathematical, it is not very clear just where the distinction comes. We next find a distinction between two mechanical systems, our ordinary 'mechanics of translation' and an unattainable 'mechanics of transformation' which would result 'if biology could ever get as close to its object as mathematics does to its own' (CE, 34); but the latter is surely not just unattainable but impossible in principle, at least if 'mechanics' is essentially a term belonging to intellect – and what else could it be?

All the time it is clear that the underlying point is the movement/trajectory distinction, but much less clear is just how it applies. The original point was that science could capture trajectories but not movements. Now it looks as though only irreversible movements, movements that involve novelty, are excluded; but I doubt if the point is really different: (Cf. chapter 6 §11 above). A further feature attributed to evolutionary change is that it is unique and unrepeatable. Evolution only happens once, and cannot be duplicated in our laboratories. This might suggest a mere epistemological point, that evolution is not amenable to the rigours of experiment, but the real point is that because evolution is irreversible the initial conditions could never be repeated. One might still wonder why it couldn't be repeated in another specimen. If, say, a giraffe evolves a long neck maybe its descendants couldn't lose it and evolve it again, if evolution is irreversible, but why shouldn't another giraffe evolve it elsewhere? Perhaps the process would not be *exactly* the same, but that could apply to almost any change – as indeed Bergson insists (CE, 48).

More surprising and more significant than his rejection of mechanism is Bergson's rejection of finalism. It commits the same sin, for it 'implies that things and beings merely realize a programme previously arranged'; it excludes invention, creation, and the unforeseen, and 'time is useless again' (CE, 41). It is 'inverted mechanism', which substitutes the attraction of the future for the impulsion of the past' (CE, 41, 42). This sounds as though finalism involved a sort of backwards causation, though one might think that if something moves towards a predetermined end this leaves room, in a way unknown to mechanism, for spontaneity in the selection of means to the end. Bergson rejects 'internal' finalism, i.e. that version which says that each individual is guided by its own internal ends. Individuals are not sufficiently individual for this, and it would lead to paradox when one individual formed part of another. He regards this, incidentally, as a form of vitalism, though his own theory is often called by that name.

4.Scientific arguments

When he turns to the scientific arguments Bergson starts by proposing that mechanism will be refuted if we find that 'life may manufacture the like apparatus, by unlike means, on divergent lines of evolution' (CE, 58). For how likely is it 'that, by two entirely different series of accidents being added together, two entirely different evolutions will arrive at similar results?' (CE, 57). Yet these results are what we observe in the case, e.g., of the eye, an extremely complex organ which is remarkably similar in the case of vertebrates and certain molluscs, which are agreed to have had no common ancestor since before this development started. Bergson eloquently states the problem:

> How could the same small variations, incalculable in number, have ever occurred in the same order on two independent lines of evolution, if they were purely accidental? And how could they have been preserved by selection and accumulated in both cases, the same in the same order, when each of them, taken separately, was of no use? (CE, 68)

The example is far too complex to discuss in detail here, even if I were competent to do so. The phenomenon is known as that of analogy, where the same function is performed in different creatures by biologically unrelated structures, as against homology, where similar structures have developed from a common origin, irrespective of present function. (See Penguin *Dictionary of Biology*, 7th edition.) Analogy is often ascribed to adaptation, but here Bergson makes a good point: 'adaptation' is ambiguous (CE, 61). (See also CE, 70, for an ambiguity in 'correlation'.) Water poured into a glass 'adapts' to the shape of the glass because it cannot help it; the glass is there to constrain it. But, if I adapt to circumstances, they don't compel me to do so; I might just perish instead. Similarly in the case of the eye we cannot appeal to any pre-existing mould like the glass to guide the development. Natural selection may eliminate unwanted variations, but what guarantees that the right variations arise? And unless each step in the development is immediately useful how does natural selection manage to 'store' it unless other changes occur which make it useful? It is not clear why storage should be such a problem unless the new abode is positively harmful, and even then storage can occur if it is recessive, as we know all too well from the history of diseases. Later writers seem more optimistic than Darwin or Bergson that the intermediate steps in the eye case are at least not harmful. (CE, 67 – 8. See Simpson 1949: 174 – 5, Dobzhansky 1955: 368 – 9.) As for so many steps occurring in the right order on different occasions, this may be less of a problem if new variants are frequent enough. Bergson thinks it surprising that two random walkers should trace curves exactly superposable on each other (CE, 60). But suppose they are walking on a mountainside with vegetation that gets progressively lusher at lower levels (lushness here corresponds to general visual efficiency), and that their policy is not entirely random but consists in walking horizontally except that every now and then they stop and sample a random selection of different paths (this corresponds to mutation and chromosomal recombination), always selecting that which leads most steeply down (this corresponds to natural selection). Then it would not be surprising that their paths, if not exactly superposable, passed through the vegetation strata in the same order. Variants arising at stages when they were not useful would be ignored as not being downhill paths. The illustration is mine, but I intend it to represent Simpson's evolutionary 'opportunism' (Simpson 1949: chapter XII). This could also be used to account for the convergence Bergson calls 'heteroblastia' (CE, 79) whereby, e.g., the same retina developed from quite different sources in man and mollusc.

I have concentrated so far on Darwinian gradualism, which in this century has formed a central plank of the 'new synthesis', along with natural selection, Mendelian genetics, and Weismann's isolated germline, later specified as Crick's 'central dogma'. This synthesis is now showing marked signs of breaking up with the emergence of 'punctuated equilibria' (Eldredge 1986), non-phylogenetic cladistics (Janvier 1984), a 'structuralist' theory with at least some kinship to traditional orthogenesis (Brooks and Wiley 1986), and even a revival of that old perennial, neo-Lamarckism (Steele *et al.* 1984). Bergson discusses versions of some of these among other views. De Vries's saltationism he thinks solves some problems but is open to objections of its own (CE, 68 – 72). (Saltationism is different from punctuated equilibrium theory: see Eldredge 1986, Dawkins 1986: especially 241.) The inheritance of acquired characteristics is empirically unsupported, and the appeal to individual effort might account for increase of size but not for increase of complexity and could hardly be applied to plants (CE, 80 – 9). Orthogenesis, at least of the mechanistic kind he attributes to Eimer, cannot provide the kind of cause that would have any explanatory value, as opposed to a mere triggering mechanism (CE, 73 – 80).

5. *The élan vital*

Bergson's own theory of evolution brings us to the second of the notions for which he is famous, alongside 'duration': the 'élan vital', usually translated 'vital impetus' but surely better known in the French. It is introduced at the end of chapter 1 and beginning of chapter 2. (See also the summary in *The Two Sources of Morality and Religion* (MR, 92 – 5).) Like De Vries's saltationism his theory makes the tendency to change inherent in species, and like Eimer's orthogenesis it gives the tendency a definite direction, but it goes beyond Eimer's theory by making the cause psychological rather than physicochemical (CE, 90 – 1). From the philosophical point of view this is one of the two main features of the theory, for it suggests something not just outside science, in the way something like the moral law might be thought to be, but competing with it and making a difference to the world. The implementation of the moral law might well make a difference to the world, but the mere existence of the law does not, unlike that of the *élan.*

Clearly this raises some big problems. If something is outside science how can it be studied by scientific methods? How indeed can it have any relation to science and the objects of science? Will not Bergson face the 'two worlds argument' that confronted Plato concerning forms and particulars and Descartes concerning mind and body: how can one relate two things that are *ex hypothesi* totally unrelated? (See Plato's *Parmenides*: 133 – 4.)

Bergson might say that it is only physics and chemistry, not biology, which cannot deal with the *élan.* But any scientific treatment of it must presumably involve measuring it, and can biology do better than physics here? He might reply that the élan is not a substance but, if anything, a force; I say 'if anything' because he seems to be referring to the *élan* when he denies the need for any 'mysterious' force to explain the separation of animals from vegetables (CE, 119). Later he does refer to the 'mysterious' way life operates, but also reckons at least partly to 'fathom' the mystery' (MR, 93). (See also MR, 254.) He might then add that our own physics accepts four so far irreducible forces, so why should there not be a fifth? It need not be a 'thing' any more than gravity is, and as with any other force we measure not it but its effects.

This is all very well – but why does Bergson insist that we cannot deal with the *élan* as with physical forces? What does he mean in fact by his insistence that it is psychological? Is it conscious? (Cf. CE, 275.) We noted a few pages back the ambivalence of his attitude to consciousness and the organic. If 'psychological' means 'conscious' then *what* is conscious? If I deliberately raise my pen we might say the pen is raised by a conscious force, but this could only mean the (physical) force was exercised by me, a conscious being. No doubt my desires etc. come into the picture and they are conscious phenomena, but only because I who have them am a conscious being: 'conscious' does not qualify 'phenomenon' and 'being' in the same way. True, he says, 'There are no things, there are only actions' (CE, 261); but we have already discussed this issue in chapter 4. If the *élan* is a conscious force this is in the sense that it is a conscious phenomenon: there must be a conscious being who exercises it. But perhaps there need not be just one such being. Perhaps the expansionism of ancient Rome was a conscious force because a lot of Romans had to act consciously for it to exist, whether or not any of them consciously had expansionism among his motives. Perhaps similarly the *élan* is a conscious force because it involves a lot of creatures acting consciously, so that no single conscious subject for the *élan* is implied. It could still have a certain unity and could develop varyingly over time, just as Roman expansionism could. But if this is what Bergson meant it seems strange if not inconsistent that he rejects the sort of vitalism that allows a separate 'vital principle' to each individual (CE, 44 – 5). Virtually always, I think, he refers to the *élan* as single (e.g. CE, 285), though in a 1908 letter he refers to *Creative Evolution* as making God the *source* (his emphasis) of the 'currents' or '*élans*' (plural) which each form a world (M, 766 – 7). It is hard to see much in *Creative Evolution* that justifies this. As he says in another letter, God only gets two or three sentences there (M, 963 – 4). The nearest is perhaps a passage where God is 'unceasing life, action, freedom' (CE, 262). (Cf. MR, 188.) But these are unusual references, and the *élan* is finite and can be obstructed (CE, 267 – 8). An alternative might be pantheism with the world as such being treated as conscious. But the few references to pantheism in Bergson are hostile, and on this view finality would be internal, albeit to the world as a whole; but so far as it exists at all he insists that it is external (CE, 43).

Only in 1914 did Bergson hear of Butler and his life-force. But in a much later letter he repudiates this apparent ally, partly for treating Lamarckian teleology as the only alternative to natural selection and partly for talking simply in images and metaphors (M, 1522 – 8, written in 1935). His own *élan* by contrast is not a mere stylistic ornament, but is meant to provide a cause for the variations which, he agreed with Butler, the Darwinians needed for natural selection to operate on but could only attribute to chance. This was indeed a problem for Darwinism, which led Darwin himself towards Lamarckism. But does Bergson give any non-vacuous alternative?

The *élan* provides a unity that random variations do not have, and leads Bergson to a sort of orthogenesis despite his rejection of Eimer. As he is ambivalent about consciousness so he is about finalism, for surely a conscious force must have ends of some sort, and so he is about impulsion: the *élan* is a '*vis a tergo*' (CE, 109), an impulsion and not an attraction; but elsewhere it is '[n]either an impulsion nor an attraction' (MR, 95). We saw earlier how he uses the movement/trajectory distinction in arguing against mechanism. It comes again when he compares nature's construction of the eye to a simple movement of the hand (CE, 97, 99). But now another feature is added: matter may resist the *élan* as a mass of iron filings might resist one's hand and take on a certain pattern in doing so; indeed the *élan* must meet resistance if life is to arise (CE, 270). But this takes us back to the two worlds argument: how can matter 'resist' a totally extraneous force? What does matter let the force do and what doesn't it? What laws are at work? It is hardly a question simply of inertial mass. Eyes are not very massive, and the *élan* is anyway not operating on any one eye, still less pushing or pulling it. The *élan*, as a *vis a tergo*, does not aim at ends, though as a result 'the first organisms sought to grow as much as possible' (CE, 104). Life meets obstacles calling for '[a]ges of effort and prodigies of subtlety' (ibid.), and the result is division (speciation?). (Cf. CE, 94.) We never learn the roles of effort and subtlety where there are no ends.

Another point on which Bergson seems to vacillate is how far the *élan* is explanatory. Eimer's orthogenesis is criticised for offering mere physicochemical triggerings which lack the explanatory force of a cause in the full sense of an impulsion (CE, 76 – 9). Presumably the *élan* will do better; but elsewhere he is less sure: mere introduction of the *élan* at least will not be explanatory (MR, 93 – 4). The waters are further muddied because explanation itself seems to be an intellectual activity inherently inapplicable to invention (CE, 173).

This brings us to the second main feature I mentioned at the start of this section. The *élan* is psychological and so its course is unforeseeable. Cunningham thinks Bergson is inconsistent in letting the present grow out of the past while refusing to let the future be implicit in the present, for the future will soon be itself present and the present past (Cunningham 1916: 126 – 30). Other writers are less troubled: the *élan*, like history, is unpredictable but intelligible in retrospect (Lindsay 1911: 240, Jankélévitch 1959: 67ff., 135). Bergson himself brings in human action here (CE, 50). (Cf. CE, 54, 55, and the strange discussion of orange at MR, 254: is the point epistemological or more than that?) The general issue of spontaneity and libertarianism arises here, which we discussed in chapter 3. It is in this retrospective sense that he allows a certain role to finalism, and accepts a certain affinity to it (CE, 42, 53 – 6).

6. *Conservation and entropy. Order and disorder*

How does Bergson's treatment of life and the world relate to conservation and entropy? The *élan* as an extraspatial force seems likely to violate them both. The conservation of energy has a conventional element for Bergson because energy exists in different forms, which are measured in different ways, and these ways, he asserts, are chosen to justify the conservation principle (CE, 255). He does not specify this further, but though there is always an element of decision in science about how to treat recalcitrant evidence, the physics of conservation not only works, as he agrees (ibid.), but evidently works better than any alternative physics (I ignore very recent challenges), which is as much as a science would normally demand. Bergson infers that the law concerns the balance between changes in relatively isolated bits of the universe like the solar system, and not the conservation of some single quantity throughout the whole universe (CE, 256). This should not affect whether the *élan* breaches the law, since the *élan* is extraspatial and so can hardly be related to systems according to their spatial positions or relations. The point seems to be that the universe as a whole is growing by the addition of new worlds and so does not really constitute a whole at all (CE, 255). The *élan* can create these worlds without violating the law as so formulated because there is in each case no containing system within which a balancing change such as the destruction of a world must occur. Let us pass over the question what this creation consists in except to say that it cannot be *ex nihilo* in the strict sense since he is going on to argue that such a '*nihil*' is a senseless notion.

Bergson says little more about conservation but turns to entropy, which is not conventional but concerns the degradation of physical changes into heat, which is ever more uniformly distributed: 'changes that are visible and heterogeneous will be more and more diluted into changes that are invisible and homogeneous' (CE, 256 – 7). This involves a decrease in 'mutability', and periods of increasing mutability, though theoretically possible, are statistically vanishingly unlikely. Since the universe is not infinite, the origin of mutability must be sought outside physics (CE, 257 – 8). The relation between life and entropy turns out to be rather ambivalent. Life cannot stop the increase of entropy but it works against it and succeeds in retarding it (CE, 259). On Bergson's own terms it seems strange that an outside force can have this halfway house effect. But his view somewhat resembles a quite different modern view where evolution is itself an entropic phenomenon wherein nevertheless the rate of entropy increase is retarded; this is because both entropy and information increase, but increase in information represents the widening gap between the actual increase in entropy and the faster rate of increase of maximum possible entropy. (See Brooks and Wiley 1986, especially chapter 2, and also their shorter though not clearer account in Brooks and Wiley 1984.)

However, Bergson's views on entropy must be seen in the context of his views on disorder and negation, which fill much of the second half of *Creative Evolution*, though he does not, I think, make the connexion very explicit. Chapter 4 opens by introducing two 'illusions'. One is that reality can be adequately described by the 'cinematographic' approach; this we discussed earlier. The other is that order is something positive that is imposed on an underlying disorder, which generates the false question 'why there is order, and not disorder, in things' (CE, 289). This is in fact discussed twice, from different points of view, in chapters 3 and 4.

The basic idea is that any arrangement of things must have some order, by definition. A total absence of order is not a possibility whose absence has to be explained, but not a possibility at all. Disorder is simply the absence of a certain kind of order, in particular a kind we might expect in the context, and its replacement by another kind. Bergson compares the way we might say of a book 'This is not verse' or 'This is not prose', but only to express its failure to meet some expectation etc.; 'I have not seen, I never shall see, an absence of verse. I have seen prose' (CE, 233). Later he adds that we perceive only a presence, not an absence: 'There is absence only for a being capable of remembering and expecting' (CE, 297). But in *Matter and Memory* there was perception too only for such beings. This view of negation resembles that criticised by Buchdahl (1961). It is developed in chapter 4 along lines suggesting Ryle's view that one can only sensibly deny that something is F by tacitly asserting that it is some other kind of G, where G is the genus of F (Ryle 1929). Bergson applies this to negative existential judgements as well as attributive ones (CE, 300 – 1, 306), but makes things easier for himself by taking the example 'A does not exist', which is hard to make sense of, if 'A' is a proper name, without attributing to A some contrasting kind of existence, e.g. fictional. (Cf. Strawson 1967.) What sort of incompleteness could Bergson claim for a sentence like 'There are no blue swans'? The nearest he comes to treating this seems to be where he hints at saying it would exclude a possible world in favour of the actual world (CE, 306). He goes on to treat negation as subjective (CE, 307); a negation is an affirmation but of second degree, presupposing another one, and is 'of a pedagogical and social nature' (CE, 304). For formal logic, 'to affirm and to deny are indeed two mutually symmetrical acts' (CE, 308), but this he thinks illusory, for effectively the same reason as in the verse/prose example above. He does not discuss examples like Frege's 'Christ is immortal', where a speech act analysis would leave it unclear whether we were affirming or denying. Perhaps excusably he does not clarify just how far he does intend a speech act analysis, where negation belongs not to what is said but to what one does in saying it. (Cf. Frege 1952.)

Bergson uses his analysis to decry the question, why is there something rather than nothing? (CE, 312; cf. CE, 290 and also CM, 71 – 7 (61 – 6).) The idea that there might be nothing is senseless, and so the existence of something needs no explanation. Intuitively we could divide this issue into three, according as we try to abolish time, space, or matter. It seems conceivable, however unscientific, that all change (including changes in our physiology and mental life) should cease, and plausible to say that time would vanish as well, though the opposite has been held by Shoemaker (1969). The absence of space seems much harder to conceive because it seems to demand more than the mere absence of matter, and empty space, despite objections from Aristotle onwards, seems conceivable enough. I cannot discuss this further except to make two points: we must distinguish the abolition of any given bit of matter from that of all matter; and we must distinguish conceiving from imagining. Bergson explicitly does both of these things (CE, 299, 295, 302). But he seems sometimes to fall into Berkeley's confusion about imagining a tree and it is not clear that he keeps consistently to the second distinction (CE, 398).

To return now to order and disorder (CE, 232 – 49): Bergson sees two kinds of order, the vital and the physical, or geometrical, as he often calls it. Disorder is simply the absence of one of these orders in favour of the other when we were expecting it. These orders are inverses of each other, and life and lifeless forces pull in opposite directions. The relations of all this to increase of entropy are not entirely clear, but since life retards but never abolishes this increase presumably it consists either in an increase in physical order at the expense of vital or in a 'degradation' of one kind of energy into another (in particular, heat), where this has the same effect as the above. (See also CE, 293, for the spin-off Bergson hopes the philosophy of the *élan* will gain from his discussion of nothingness.)

But are the two orders related as Bergson thinks? Some of his argument looks like special pleading, as when he insists that when we imagine the world as chaotic and no longer obeying physical laws we replace these laws by a multitude of wills (CE, 246). But more important is the question, What is meant by 'physical' or 'geometrical' order? If it is that order whose absence is excluded by definition it must apply to everything, living and non-living alike. But, if alternatively it has degrees and is other than life, why should there be only one such order? If it is simply *defined* as the inverse of life then it means 'ordered (in the first sense) but not alive', in which case degrees of it will simply be degrees of non-livingness, since the first sense of order does not admit of degree; are we then to understand that increase of entropy is by definition decrease of life, so that to say the *élan* can retard but not stop it is to say the *élan* can retard but not stop the decline of life? But until life gets going what is there to retard? And what happens to the law of increasing entropy in a lifeless universe, or part of the universe? And, if that law simply says that, e.g., electrical energy tends to turn into heat, why has this anything to do with ordering at all? There are certainly big problems about the nature of randomness, but it seems unlikely that they can be settled simply by appealing to order in the sense of Bergson's vital order. (For further discussion of Bergson on entropy and related issues see Čapek 1971, appendix III, and on negation see Gale 1973 – 4; cf. also Winch 1982.)

VIII

The comic

1. Introduction. Art and comedy

Exactly halfway through his life, in 1900, with two of his main books behind him and two to come, Bergson published a short book that might seem to have little to do with the rest of his philosophy. *Laughter* is perhaps the second most famous of his books, after *Creative Evolution*, and certainly the easiest to read. However removed its subject might be from his general concerns his treatment of it fits in with those concerns but without involving the technical apparatus of duration or the *élan*.

Laughter is the nearest Bergson came to writing on aesthetics in any systematic way and its third chapter contains a short discussion of art, though mainly as a foil to contrast with his main subject, the comic (L, 150ff.). (See also CM, 159 – 63 (135 – 8). *Time and Free Will* discusses aesthetic feelings, from a rather different point of view (TF, 11 – 18).) We have seen earlier how his theory of perception as it occurs in everyday life is coloured by pragmatism. We perceive what we need to perceive in order to act. We perceive objects as bearers of certain general features rather than in their full individuality. We distinguish one man from another because we need to but unless we have the special training and purposes of a shepherd we cannot distinguish one sheep or goat from another. Language too, apart from proper names, shares this feature of generality (L, 153). It is in art that we escape from this pragmatism. The artist perceives things for their own sake and so sees their inner life appearing through their outward forms, though usually, because of human limitations, only in the case of one of his senses. So 'art always aims at what is *individual*' (L, 161).

Art in fact has some relation to philosophy, though in a letter in 1915 Bergson distinguishes them (M, 1148). They both appeal to intuition, but art does so exclusively while philosophy involves intelligence too; this is because art is concerned only with the living while philosophy must concern matter as well as spirit. Also philosophical intuition goes far beyond artistic intuition because it seizes life before its scattering into images, while art deals with these images. One wonders how he would treat, say, the paintings of Mondriaan. And consider a neat and simple mathematical proof, or the organised complexity of a detective story, or a Bach fugue: their invention for Bergson would involve intuition, but the aesthetic appreciation of them surely involves intelligence too. But a treatment of art which is subordinated to one of the comic, linked to it via tragedy and drama in general, no doubt cannot be expected to be complete or to answer all questions; and in the context Bergson is primarily concerned with the depiction and appreciation of character and it is insight rather than intellect that divines the finer points for us in that sphere. It is also plausible to contrast the aesthetic with the pragmatic, in so far as we appreciate art for its own sake and not to help us manipulate the world. But will the account cover functional art, where appreciating the object's aesthetic value essentially involves appreciating its fitness to serve a pragmatic purpose? It is true that the pragmatic purpose need not be our own, so that we can be disinterested. But where does the perception of individuality come in when we are appreciating an object's fitness to serve a purpose?

Perhaps these objections could be met. But, though Bergson wants his theory of art in general at least to cohere with and give some support to his philosophy as a whole, if not to be strictly derived from it, and though he applies his theory to art in general rather than to dramatic art (L, 157), it is dramatic art that he has primarily in mind, and he points out rightly enough that Hamlet and Othello are unique individuals: one could not imagine *Othello* being re-titled *The Jealous Man*, though such titles are common in comedy (L, 15). *Othello* is a good example for Bergson to take, for it is as hard a case as one could find for this feature of his theory. Whatever else he is Othello is jealous and this is no doubt the first thing one thinks of on hearing his name. But Othello does not 'typify' jealousy; we don't say of someone, 'He's an Othello'. Perhaps we might say, 'He's a Shylock' – but probably not unless all we know of Shylock is that he demanded a pound of flesh.

In comedy it is quite different, so Bergson avers. We can indeed, in France at any rate, as he assures us, say of someone, 'He's a Tartuffe' (L, 163). Actually this again is a hard case for Bergson since the play in which Tartuffe appears carries his name as its title, but many of Molière's other plays have type-names as titles, like *L'Avare* or *Le Misanthrope*. The point is that the heroes of comedy are types rather than individuals, at any rate when we are talking of comedies of character. But it is not clear that Bergson is on such solid ground here as in what he says about tragedy. Unless we trivialise the issue by suitably limiting the phrase 'comedy of character' counterexamples seem to confront us: is Falstaff simply a type, or Figaro, or even Don Quixote, of whom Bergson makes much use?

Not all comedy is of the same kind and Bergson is careful to distinguish comedy of character from comedy of situation as well as from other kinds. Nor is it easy to decide what counts as comedy at all. To avoid begging questions it is probably best to define comedy extensionally as including those items that are generally agreed to be comedies. But the immediate question is whether something can be both a comedy and concerned to depict an individual who is not just a representative of a type. In the absence of a theory such as Bergson's, it is not clear a priori why not, and I have just suggested one or two examples. But the issue may well depend on whether, when presented with a theory, we are sufficiently impressed by it to push it towards becoming analytic, much as in science whales are not a counterexample to a scientific account of fish, because 'fish' is no longer defined simply as 'sea creature'. We must see how far Bergson's theory inclines us in this direction.

2. *The basis of the theory*

We have entered on our discussion of Bergson's theory of laughter and the comic via his theory of art, since this is the field of philosophy in which it occurs; this has led us to the comedy of character, which he regarded as the culmination of the comic (it contains 'the loftiest manifestations of the comic' (L, 132)), and which he reserved for the last of the three chapters that form *Laughter*. There also his discussion of art occurs, on which he hoped the discussion of comedy would throw light (L, 131). We have therefore entered *Laughter* by the back door, but we must now go round to the front to see how the theory is built up and what its basis is.

Bergson explains his procedure and aims in the preface and appendix which he added to the 1924 edition of *Laughter*. (These are not in the translation; see Q, 383, Q, 483, and also Q, 1504.) Other writers had tried to enclose comic effects in a broad and simple formula, but had at best provided necessary conditions but not sufficient conditions: one could always apply their formulas and come up with something that was not comic. His own aim was to provide sufficient conditions even though they might appear 'at first sight' not to be necessary, because of a certain complication we will come back to near the end. He wanted to provide 'recipes for manufacturing' the comic ('procédés de fabrication'), and he did so in terms of the contrast his own philosophy involved between life and the lifeless, though he goes further and in effect replaces 'living' by 'human': we might laugh at an animal, but only because we thought of it in some way by comparison to a human – this would come under the 'complication' I mentioned a moment ago. Man has been defined as 'an animal which laughs', but could equally well be defined as 'an animal which is laughed at' (L, 3 – 4). The general point is that we laugh at situations where the living, and in particular the human, behaves as though it were non-living.

So far we would expect the archetypal humorous situation to be where a man slips on a banana-skin, and this is just what Bergson gives us as his first example: 'A man, running along the street, stumbles and falls; the passers-by burst out laughing' (L, 8). This rather neolithic scene requires some filling out. First it is essential that the stumbler does not hurt himself, for if he does he will (we hope) arouse pity and pity is an emotion; but laughter 'has no greater foe than emotion' (L, 4). (Cf. L, 139, 185.) Laughter is a function of intelligence, as anyone confining laughter to humans would presumably agree. Bergson assumes that it must therefore involve an absence of feeling, but even if it does, this hardly *follows*. A good deal of emotion is compatible with the exercise of intelligence, as when we need all our wits to escape from danger; extreme fear may paralyse us, but some may be needed to stimulate our inventiveness. Bergson surely goes too far when he says the comic can only affect us when it falls 'so to say, on the surface of a soul that is thoroughly calm and unruffled' (L, 4). Some emotions and some degrees of emotion are inconsistent with laughter. Others fit well enough with it, and not simply because they are caused along with it by the comic. Nor is it exclusively the pleasant emotions that can go with laughter. Humour can be bitter and sardonic, and hysterical laughter, though obviously a deviant case, merges into the ordinary kind; also laughter can conquer emotion rather than vice versa, as when we amuse a suffering child. Does Bergson mean that we cannot laugh at the source of emotion, except perhaps when overcoming the emotion? But even this is not entirely true. We can laugh at something pleasing, and also appreciate a joke at our own expense.

Another feature of laughter Bergson insists on is its social nature: we laugh when in company and often do not laugh at something others are laughing at if we are not part of the group (L, 5 – 8). There is something in this, especially if he is thinking of literal laughter as against mere comic appreciation; but the reason he gives makes it depend on the socially corrective function of laughter, which we will come to later, and it is not obvious that it does so depend.

Laughter then is essentially human and involves the intellect. Bergson makes no attempt to connect it to intuition, even though life is the peculiar sphere of intuition while intellect deals with the lifeless. The comic connects the two in that it involves the living behaving like the lifeless, but it is hard to decide whether intuition should therefore be required for the appreciation of it. However, one could argue, though Bergson doesn't explicitly, that the behaviour which is laughter's target is that of instinct, and that intuition is less suited than intelligence to contrast with this.

The target of laughter is clumsiness, rigidity, the automatic, or 'distraction' (which *Laughter* translates as 'absentmindedness', though Bergson has more than absentminded professors in mind). Why we laugh at the stumbler is because of 'the involuntary element in this change, – his clumsiness in fact' (L, 9). But these are not quite the same thing. He is clumsy because he should have avoided what tripped him up, but instead went on running as before, 'automatically', when that was no longer appropriate. But what if he was tripped by something he could not possibly have avoided – as perhaps happens in slapstick like the Tom and Jerry cartoons, though it is not altogether clear how far we think of such characters as 'able' to avoid their misfortunes, or how far this matters? Not even Bergson presumably would laugh at someone thrown flat on his back by an earthquake. His view of the function of laughter again suggests that the comic is not simply the spectacle of something rational behaving as an inert physical object, but that of its so behaving through its own fault and because of a defect of rationality imputable to it.

One methodological point must be made before we go on. Bergson criticises other views because they only give necessary conditions for the comic and not sufficient conditions which could provide a recipe. Laughter is often explained, for instance, in terms of surprise or contrast: we laugh at things because they are unexpected or not as they should be. But this clearly won't do because there are so many surprises or contrasts that we don't laugh at (L, 39 – 40). But does Bergson escape the same criticism?

We laugh at human automata – but we don't if our emotions are aroused or if we are not in company. If Bergson's theory can be tidied up by exception clauses why couldn't other theories too? He might reply that they have not yet been so tidied up, but the point is that the mere fact that we do not always laugh at contrasts etc. cannot be fatal to the idea that they form the basis of the comic if the same objection is not fatal to the automaton theory.

3. Development of the theory

So far we have only the basis of the theory, yielding the crudest form of humour. Bergson goes on to develop the theory to cover more complex kinds, including wit and verbal humour and leading up to the comedy of character from which we began. How far will the theory bear these extensions?

The extensions really come in two stages. The first gives various rules which spell out what the rigidity or automatism we have mentioned amounts to in various cases. When discussing the comic in situations Bergson develops at some length three such rules, involving repetition, inversion, and what he calls 'reciprocal interference of series' (L, 89). If something is repeated again and again this suggests an inanimate process, for repetition is alien to the sphere of life and duration, as we saw in earlier chapters. Bergson takes examples from Molière in particular where a certain pattern is repeated with different sets of characters and possibly with other differences. This is something that would be out of place in a tragedy, though one might wonder whether it would in a tale on a theme of being haunted. However, it is not simply an application of the rule to make your comic characters behave as things, since no actual character is required to repeat his behaviour. It is rather that one character repeats the behaviour of another, and without intending to do so, at that: he is not a comic character because he follows in a wooden fashion what someone else does; the repetition is at a more abstract level, and presumably can only work, on Bergson's theory, because the spectator latches on to the idea of repetition, ignoring that repetitiveness is not imputable as a character trait to the comic personage in question.

With inversion 'the root idea involves an inversion of *roles*, and a situation which recoils on the head of its author' (L, 95). We can accept the comic possibilities of this, but Bergson does not say much to distinguish it from the tragic peripeteia of an Oedipus. The connexion with the automaton motif is that as living phenomena do not repeat themselves so too they do not go into reverse (L, 89).

The third notion is harder to grasp or to explain briefly, but involves the linking of two independent series of events by events which can be ambiguously interpreted as belonging to either of them, thus compromising their independence by suggesting they consist of interchangeable parts in a way no living phenomena would (L, 101). Bergson gives one example, from Daudet, and refers to the mock-heroic and the 'transposition from the ancient to the modern' (L, 100). It is not clear how widely he is ranging here but surely too widely if he would include Shakespeare in modern dress or O'Neill's *Mourning Becomes Electra.*

One general comment on the automaton motif, in view of Bergson's claim to provide a sufficient condition or recipe for comedy, is this: could one not depict rigidity for didactic purposes, as an 'awful warning', without any comic effect, and could not inflexibility be at least part of the flaw of a tragic hero? Perhaps Bergson would claim that our emotions would be involved, but what determines that the situation should cause emotion rather than mirth?

Most comedy above the banana-skin level is expressed in words, but some seems essentially connected to them, and Bergson devotes the second half of chapter 2 to 'the comic in words'. He starts by distinguishing the comic created by language from that merely expressed by it, the criterion being that the former is usually untranslatable, but he doesn't in fact confine himself to the former, for many of the ensuing examples are perfectly translatable. But for those which are not 'it is language itself that becomes comic' (L, 104). This leads him to a certain embarrassment, for he finds it hard to say who we are laughing at in such cases. He decides that when a word is comic we laugh at its utterer and when it is witty we laugh at either a third party or ourselves (L, 104). It is significant for the nature of his whole theory that we must be laughing at some*one* and not simply at a joke. But how plausible is this? To make it so he bases the witty on the comic: a witty remark encapsulates a comic scene. His example is, 'Your chest hurts me', written by Mme de Sévigné to her ailing daughter. This suggests to Bergson a scene in Molière where a doctor diagnoses someone's illness by taking her parent's pulses, which in turn suggests the marionette motif, father and daughter sharing a pulse as though joined by invisible strings. But are we laughing *at* Mme de Sévigné, or her daughter, or both, or at ourselves for taking the remark as we do? Why any of these?

Another source of humour Bergson mentions involves the literal interpretation of metaphors. 'Every word, indeed,' he claims, 'begins by denoting a concrete object or a material action', abstract senses coming later (L, 115). A recipe for the comic is therefore to pretend to take literally words intended figuratively. The connexion with his general theory is evidently that something at the material level is being substituted for something at another level, though that level is rather the abstract than the living. Perhaps the point is that someone who so misinterprets what is said is behaving as though he only understood the literal sense, and so is, as it were, on a lower plane, though hardly himself just a material object or even an animal. But if this is what we are laughing at why is it we laugh more when the misinterpretation is only a pretence? A real misinterpretation would hardly elicit more than a derisory snort; it seems to be the skill of the comedian that elicits real laughter. Incidentally at one point he presents as equivalent two formulations of the relevant recipe, but only one of them refers to pretence (L, 115).

The rest of the chapter applies to this sphere the notions we met above of repetition, inversion, and reciprocal interference. The last covers puns, and play on words, which is amusing because it 'always betrays a momentary *lapse of attention* in language' (L, 121). Again we might wonder whether it is the lapse itself that generates real wit as opposed to a low-level snort, though Bergson may be right to the extent that *some* source of humour must lie in the material if we are to admire the humorist's skill in searching it out, and if this skill is itself to contribute to the status of the result as wit rather than something lower. Of the three notions Bergson thinks repetition the most important, but it now takes the form of transposition of level, as when the solemn is treated as frivolous or vice versa; but though this may *involve* repetition, because the levels must have something in common, the comic effect surely cannot be reduced to this, and to that extent has not been derived from the mechanical.

But are we perhaps ourselves being too rigid in our attitude to Bergson's theory? Here we reach the second of the two stages I mentioned at the start of §3, and also the complication mentioned near the start of §2. In the last paragraph of chapter 1 he insists that his theory has a certain open texture, as we might say, in that many forms of the comic can only be analysed by their resemblance to the central form we have been discussing; the case is somewhat, though not exactly, like that of Aristotelian focal meaning. As he says in the appendix, we laugh in cases which have a superficial resemblance or accidental relation to the central case or to cases themselves so related to it (Q, 484). The circle can expand indefinitely and the reason is that we enjoy laughing. (See L, 102.) Writing to a reviewer of *Laughter* he said that he had exaggerated the rigidity of his rules for clarity of exposition, and that the correct analysis of any given case might be far from obvious and the thread leading to it extremely thin (M, 436 – 7). He hints at the *élan* as causing the comic 'to glide thus from image to image, farther and farther away from the starting-point, until it is broken up and lost in infinitely remote analogy' (L, 65), without making very clear whether this gliding is a process that occurs in time, though it could be if we think of the peripheral cases as being the more sophisticated and historically later to develop. (See L, 2, 95, 187.)

All this tends to disarm criticism, for who could dispute an 'infinitely remote analogy'? But we can conclude that Bergson has provided a suggestive and valuable theory which does throw light on a good many cases, whether or not it goes as far as he wants it to.

A particular feature of Bergson's approach is that he ascribes a function to laughter. 'The comic is that side of a person which reveals his likeness to a thing' (L, 87). It therefore 'expresses an individual or collective imperfection which calls for an immediate corrective. This corrective is laughter, a social gesture that singles out and represses a special kind of absentmindedness in men and in events' (L, 87 – 8). (Cf. L, 134, 136, 170, 174, 197 – 8.) Perhaps laughter grows out of states which did have a function, whether the smile that soothes a baby or the braying that intimidates an enemy, but what reason is there to think it must have a function now? Is it really true that in laughter 'we always find an unavowed intention to humiliate' (L, 136)? Bergson has to admit that so far as the justice of its picking of targets goes laughter is a blunt instrument (L, 197 – 8). He implies that this is because laughter is spontaneous and has no time to consider. But does he mean that we often laugh at people when we shouldn't (banana-skin victims), or that established humorous episodes, such as the literary ones he quotes, are often such that we ought not to laugh at them really, though we continue to? In that case it is hardly a question of time. In fact at one point he seems to confuse the role of laughter with its target, when he says that comic observation goes to what is general so that the correction shall reach as many people as possible (L, 170); but usually, as we saw earlier, it is because of the repetitive, automatic, nature of the physical world to which comedy is akin that comic characters are types rather than individuals. While on the view he takes of the comic laughter could have the function he assigns to it, he gives no real reason for saying that it does have it. But let me end with his own summary sentence after he has noted that humour delights in concrete terms, technical details, etc.: 'A humorist is a moralist disguised as a scientist' (L, 128).

IX

Morality and mysticism

1. Introduction

Bergson's last main book was published during his arthritic old age in 1932. *The Two Sources of Morality and Religion* is as much about anthropology and sociology as about morality or religion, but it is by far his most serious attempt to write on morality in particular. Back in 1912 he had expressed doubts about whether he would ever publish a book on morals, because he doubted if he could ever reach conclusions as demonstrable or as 'presentable' ('aussi démontrables ou aussi "montrables"') as those of his other works, for philosophy proceeds by a definite method and can thereby claim as much objectivity as the positive sciences, though of another nature. Three years later he refers to himself as having been occupied with moral problems for several years, and in both places he refers to such problems as inseparable from or presupposed by problems about the nature and creative activity of God (M, 964, 1147).

The idea that philosophy can rival science in its demonstrability, though a claim Bergson entertained seriously, can perhaps be put down to professional optimism, but his method, as we shall see, does have much in common with that of his earlier works in so far as he applies some of the main ideas he had developed there, perhaps altering them to some extent in the process. He makes much use of the idea of increasingly divergent processes which are reconciled or transcended by some kind of synthesis, and he also constantly appeals to the 'Zeno point', as we may call it, that a real process can never be constructed out of the atoms into which its trajectory can be divided – though now he applies it to historical developments in the world rather than to theoretical constructions by us.

But the mixing of ethics with anthropology leaves room for some uncertainty about just what Bergson is trying to do, an uncertainty reflected in the very title of the book; for what does he mean by the 'sources' of morality and religion? The word suggests origins, in the historical or evolutionary sense, especially when applied to religion. But it could also refer to the logical origins, or justification of something. What he in fact gives us, for both morality and religion, is an account of two *types* of each of them, which he calls closed and open in morality, static and dynamic in religion, and, though he does indeed talk about the historical or psychosocial origins of these, it is types rather than origins that there are really meant to be two of. Most of the book, as we would expect, is concerned to describe and distinguish the two types in each case, and Bergson clearly regards one of them, the 'open' and the 'dynamic', as superior to the other, the 'closed' and the 'static'.

The book has not very much ethical argumentation, properly so called. But a good deal is presupposed, even if it is not argued for or is defended only by implication. Much in fact depends here on how far the evolutionary discussion is teleological in nature. In *Creative Evolution*, as we saw, Bergson rejects finalism along with mechanism, of which it is a sort of inverted version, a *vis a fronte* rather than a *vis a tergo* but just as rigid and determinate as mechanism. Whether this is still the case in *Morality and Religion* we shall have to see. One who believes that there is an overall purpose in the universe is not committed by any logical principle to regard that purpose as a good one. But it seems that virtually everyone who has held such a view has in fact regarded the purpose as good, provided at least that it is the only or the dominant purpose. Even Manicheans believe that good will win out in the end. If therefore the *élan*, or whatever is responsible for implementing the purpose, can be shown to produce or tend towards producing a certain moral outlook it is easy to see that this fact alone will often be regarded as justifying that outlook. The evolutionary ethics of someone like Spencer provides an obvious example. Bergson is less explicit in praising the evolutionarily advanced as such. But his position might be put like this: human nature being what it is, and the material the *élan* has to work on being what it is, it is inevitable, or at any rate only to be expected, that moral ideas should emerge in the way they have, particularly in the case of the closed morality. The open morality is superior to the closed, in Bergson's view. But is not Bergson himself a product of evolution? The questions that loom at this point are of a kind he does not discuss at all. He would no doubt claim that he is not a determinist, and so need not regard his own or anyone else's views as the product of a causal process which cannot, on pain of over-determination, be the product of a rational one.

The implied contrast here is disputable: why should not the causal process involve steps which are drawings of rational inferences? If this is so, neither Bergson nor the determinist need be worried by this particular argument. Yet there will be a tension in his position if he is trying to provide a Hume-style 'naturalised ethics' whereby the advance of ethical ideas is to be explained as due to an evolutionary process whose own nature as purposive determines it to produce good or correct ideas, judged to be so by us, the products of that same process, who also judge that the process is likely to produce good results. It is a complication that the process is presumably not yet ended, so that the results so far, together with judgements on them, may not yet be correct; but they will be better than their predecessors, and so we can ignore this complication, unless someone insists on the rather abstruse objection that the present epoch could embody a backsliding which, however prolonged, was relatively minor by comparison to the whole process, with its unknown length – but let us ignore this objection too.

The tension arises like this. As I argued above, it is not impossible for our views to be both causally produced and correct. But, if we regard them as causally produced, must it not occur to us that had the causal process gone differently we might have had quite different ideas, and regarded both them and the process which produced them as correct and properly purposive? How then can we feel any assurance that the process which in fact produced them did so *because* it was a good-seeking process, likely to produce correct results? Must we not regard the process as self-adjudicating?

So much then for the general nature of the task Bergson sets himself. Let us now turn to some of the actual content of what he says. Of the book's four chapters the first is devoted to morality as a whole, the next two to the two types of religion, and the fourth to certain features of modern industrial civilisation and some remedies for its discontents.

2. *Obligation*

Moralities are often divided into teleological and deontological. In the former the end is paramount and our obligations are governed by the need to achieve the end. In the latter the obligation comes first and value is constituted, in part if not in whole, by the fact that obligation is fulfilled. We have seen that Bergson's attitude to finalism is somewhat ambivalent, but one might think there was enough of the teleologist in him for his approach to ethics to lie in that quarter. He does not make the contrast as such, but though he is far from being a Kantian he gives much more attention to the notion of obligation than to that of goodness. To this limited extent he might be called a deontologist, with teleology coming in when we ask about the content of our obligations, which is linked, as one might expect on an evolutionary view, to social cohesiveness. But there are at least two complications. One concerns the distinction between closed and open moralities, to be discussed later; at this stage we are concerned with the closed. The other concerns the tension I have already discussed between simply describing the growth of moral ideas on the one hand and subscribing to and trying to underpin those ideas on the other.

Obligation is rooted in habit and ultimately in instinct, the social instinct of the social animals. It exists in us only because as creatures possessed of intellect we are tempted to rebel and resist the automatic behaviour instinct would elicit from us; we then resist this resistance, and this second-order resistance takes the form of obligation (MR, 12). It is the call of the normal, the regular, the lawlike, pulling us back from deviancy; for though the philosophers insist on distinguishing laws of nature from prescriptive laws, to ordinary men the distinction is less clear: 'A law, be it physical, social or moral – every law – is in their eyes a command' (MR, 4). But obligation is not some strange force imposed upon us from outside; it exists only because of this pull of instinct within us, whether or not this takes a social form – so at least I interpret the difficult discussion where we are told: 'Obligation, which we look upon as a bond between men, first binds us to ourselves' (MR, 6).

But we mustn't think of each or even any particular obligation as instinctive, for we need intelligence to tell us what our obligations are. Obligation in general is 'the very form assumed by necessity in the realm of life, when it demands for the accomplishment of certain ends, intelligence, choice, and therefore liberty' (MR, 19). It is a 'virtual instinct' (MR, 18), and Bergson compares the way speech has an instinctive basis, though all the actual elements of language are conventional and not natural.

Obligations then may be given to us by reason, but obligation is not. One might perhaps compare the view of conscience taken by P. Fuss (1963 – 4): conscience does not tell us what we ought to do but tells us to do what we already think we ought to do; it is motivational rather than cognitive. Similarly for Bergson obligation is the force that binds us to what we would do by instinct if we were purely instinctive creatures, but, since we are not, intelligence is needed to specify our obligations.

Bergson sees a certain, very limited, similarity between obligation and the Kantian categorical imperative, though he gives a peculiar interpretation to that imperative, as though it said simply 'You must because you must' and were typified by a military order (MR, 15). He imagines an ant which momentarily rebels against its self-sacrificing instincts but is immediately drawn back to them without getting as far as formulating reasons why it should be (MR, 15 – 16). This apparent travesty of Kant's real view probably depends on Bergson's wanting simply to emphasise the 'categorical' element in the imperative without discussing its basis in reason. He discusses Kant again later, though without much sympathy (MR, 69 – 70). But there is another element in Kant's thought that he rejects, namely that obligation is always felt as an alien intrusion on to our inclinations; this involves, he thinks, confusing 'the sense of obligation, a tranquil state akin to inclination' with the violent effort we sometimes need to break down an obstacle to it (MR, 11). At first sight this sits uneasily with the idea that obligation only arises when we rebel against our social instincts. But one thing he insists on is that the great majority of things we regard as our duty we do as a matter of course and without reflection. It is only occasionally that we feel a real moral conflict. There are perhaps two points to make here to ensure that his position is consistent: first, a rebellion can exist without being anything but a very mild one; and, second, he could say the sense of obligation represents a mere intellectual apprehension of a gap between what our inclinations and our social instincts would lead us to do, a gap no sooner seen than closed as our erstwhile inclinations are superseded by the motivating force of the social instinct.

3. The question of justification

This then is how a sense of obligation grows up for Bergson. But can it be justified? Is he simply presenting us with morality as a going concern, take it or leave it, and with no answer to the sceptic who prefers to leave it? One might think of Hume's reply to this question at the start of the second *Enquiry*: he denies the genuine existence of such a sceptic. Once again Bergson has something in common with Hume, though Hume was not one of his favourite philosophers. (Hume has not a single reference in the index to *Oeuvres*. Of the six references in *Mélanges* only one, a hostile one on causation, is anything but purely casual.) Bergson would say, I think, that the sceptic will simply find the social instinct present in himself, and cannot 'leave it', except verbally. Selfishness of course exists.

> Yet, if utilitarian ethics persists . . . this means that it is not untenable . . . because, beneath the intelligent activity, forced in fact to choose between its own interests and those of others, there lies a substratum of instinctive activity, originally implanted there by nature, where the individual and the social are well-nigh indistinguishable. (MR, 26)

Later he shows in more detail how egoism is driven to become enlightened egoism, along lines familiar enough (MR, 71ff.). The inadequacy of this approach as a justification for morality is also familiar, and he fails to distinguish the other-interested from the other-regarding, but this does not matter, since he goes on immediately to realise that such considerations can only give us hypothetical imperatives (MR, 73). (Cf. MR, 26.)

Is there then no reason to justify morality because its existence for us as a source of motivation is inevitable? Bergson's language is ambiguous. He insists that obligation exists 'already' (MR, 76), and after showing the failure of one justification he goes on: 'Yet we *are* obliged, and intelligence is well aware of it, since this is the very reason why it attempted the demonstration' (MR, 75). Are we obliged, or is it only that we feel ourselves to be so? Or is there for Bergson simply no difference between these two, allowing perhaps for correction of our feelings sometimes by a more sensitive attunement to social pressure? We have in fact reached the 'naturalised ethics' I mentioned earlier, and he sums it up when he calls 'any theorising on obligation' 'unnecessary as well as futile: unnecessary because obligation is a necessity of life' (MR, 77). It is futile because intelligence can only justify incompletely, even to itself, an 'anterior' obligation.

We have already seen that reason may give us obligations though not obligation, and no doubt it will also play a part in the 'attunement' I have just mentioned. But when Bergson comes to his frontal attack on the possibility of justifying morality by reason he starts from yet another point of kinship with Hume, that reason can only be the slave of the passions: 'if [reason] spoke only in its own name . . . how could it struggle against passion and self-interest?' (MR, 69) Obligation cannot 'originate in bare ideas' (MR, 81), 'as if an idea could ever categorically demand its own realization!' (MR, 78).

The attack itself has two prongs, according as reason is taken as purely formal and demanding mere consistency, or as having content and prescribing a certain end (MR, 69ff.). The first prong attacks Kant's treatment of the duty to return a deposit on the grounds that not to return it would be to treat it as not a deposit. (See Kant 1873.) This, says Bergson, either juggles with words or begs the question by assuming the obligation is already there. This brusque dismissal does little justice to Kant's attempt, unsuccessful as it may be, to show that reason contradicts itself in willing for the agent what it cannot will universally.

Before passing on Bergson takes time off to reject any attempt to base morality on a Platonic Idea of the Good, a rejection which merits just two comments. First, this is, I think, the nearest he comes to an explicit treatment of a Moorean teleological ethics, though what he says of it hardly goes beyond what Aristotle said of Plato (*Nicomachean Ethics* I 6). And, second, he makes the interesting comment that the reason the Greeks were silent about duty as we know it was that they took it for granted: morality only intervened to grace the philosopher's life in a quasi-aesthetic way *after* he had done the duties his city demanded.

The second prong, concerning whether reason can supply a content for our moral activity, is equally disappointing. The main burthen is that reason would simply presuppose certain ends as given from elsewhere, but it seems that a threefold confusion infects his thought. Intelligence will

> persuade itself into thinking that an intelligent egoism must allow all other egoisms their share [and] will build up a theory of ethics in which the interpenetration of personal and general interests will be demonstrated, and where obligation will be brought back to the necessity, realized and felt, of thinking of others, if we wish intelligently to do good to ourselves.
>
> (MR, 75)

Are we looking for a demonstration that altruism is rational, a demonstration that it is right, or a demonstration that will somehow make us altruistic? The relation between the first two of these would return us to the first prong, but it seems that nothing less than the third would really satisfy Bergson, and this he thinks would be impossible because of the Humean wedge he drives between reason and motivation.

Some further light is thrown on Bergson's real position by a discussion near the end of *Morality and Religion*, where he claims that generally accepted morality can be deduced from a whole variety of principles: 'General interest, personal interest, self-love, sympathy, pity, logical consistency, etc.' (MR, 232). The trouble is, he thinks, that the deduction is both too easy and only approximate, and the philosopher in fact finds what he puts there, but leaves 'morality itself' unexplained through want of delving into 'social life' and 'nature herself', which a 'purely intellectualist philosophy' does not do'; it rather seeks refuge in a Platonic theory of the Good, which cannot explain 'how the ideal in question creates an imperative obligation', for 'an ideal cannot become obligatory unless it is already active' (MR, 232 – 3).

The attack on intellectualism evidently derives from the view that reason is inert, but it is now more general, recalling the limitations of intellect that we have seen from previous works, and it goes on to bring in the Zeno point, so common in *Morality and Religion*, that you cannot get from a trajectory to a movement by a smooth transition. It is also set in a context where the two moralities, closed and open, are under discussion, and it is worth noting that the Zeno point has a double application in ethics. Here it applies to both moralities together and concerns the impossibility of generating a motivational force from purely theoretical ideas (in this application it could be called the Hume point). The precise application seems to come to this: the points on the trajectory correspond to stages in an intellectual argument, while the movement corresponds to the psychological process of actually being motivated by the argument as a whole. It is unclear that the comparison will stand up to detailed examination: why should it not apply as much to intellectual acceptance of the argument as a whole, or indeed to intellectual comprehension of it, as to becoming motivated by it? But the point is evidently meant to be that motivation is something different in kind from intellectual comprehension or acceptance. Whatever we may think of the Hume point as such, we may see an element of merit in Bergson's position if we ask what is our own attitude to the moral status of people who understand and accept moral arguments on a purely intellectual level but are genuinely not motivated by them at all. Such people seem to exist, in the form of psychopaths, and whether or not we allow that they really do accept the arguments in any sense at all it is at least plausible to claim that they cannot strictly be blamed for not acting on them.

The second application of the Zeno point, as we shall see, concerns the transition from the closed morality to the open, and we must now turn to this distinction.

4. Closed and open morality

One thing that is closed about the closed morality is that it is not universal in its scope. It is a morality of group-loyalty, which arose in, or rather existed from the beginning in, the earliest human societies (MR, 233 – 4). There are problems about the evolution of altruism, but it seems unproblematic that a species whose members were willing to sacrifice themselves for their already born offspring would be evolutionarily favoured, which gives us at least a start. (See also Ruse 1986.) The main problem in Bergson's eyes is that group-loyalty can never turn into something else, such as universal benevolence or the brotherhood of man, by mere expansion of the group, because it essentially needs an out-group as a foil. This is a much-used point. Perhaps humanity will only finally unite when we can vent our hatred on the Martians. But the point has its limits. It presupposes that love of our fellows must rest upon loyalty and cannot come from a realisation that there is an identity of nature between ourselves and our friends on the one hand and any man as such on the other – but here we are running up against the Hume point, and later Bergson adds that the sharing of all men in a 'higher essence' is not obvious (MR, 199 – 200). Even Bergson, however, should surely consider the idea that though an out-group may be necessary it need not be human or even animate. Could not mankind unite against not quasi-human Martians but an invading disease or an ice age?

However this may be, it is the impossibility of getting from the closed to the open by expansion that leads Bergson to the second application of the Zeno point. (See e.g. MR, 45, 46, 57 – 8; he seems to conflate the first and second applications at MR, 25 – 6). The open morality differs from the closed in kind, not in degree, though once the transition has been achieved the effect in retrospect is as though all the intermediate degrees of broadening had been gone through, and the closed morality, 'enclosed and materialized in ready-made rules', is like a 'snapshot' of the other (MR, 46).

Two questions arise then: how *does* the transition occur, if not by broadening, and what is this end-state that differs in kind from the starting-point? The two questions can hardly be separated. The end-state cannot be just love of humanity (though in a way it involves this), because it is not the same feeling that one has for one's family etc. only with a wider object. It is, however, very much a state that depends on feeling or emotion, for this is what gives the impulse that replaces obligation (MR, 25). But the emotion is of one kind and not another, not the kind which follows on intellectual representation which it has had no part in causing, but the kind which is 'pregnant with representations' (MR, 32), and is supra-intellectual rather than infra-intellectual. (It is here alone that women, in general, fall below the level of men, whose equals they are in intelligence (MR, 32 – 3).) This power of the emotion to generate intellectual representations and moral rules means, Bergson claims, that we have not here an eighteenth century sentimentalism (MR, 35) – though one might wonder why such sentimentalism could not itself generate moral rules, if we mean 'lead to' and not 'validate'.

One might expect some reference to intuition in all this, and it is indeed there in the background, surfacing several times in the later chapters, though in the first chapter it only makes one casual appearance (MR, 34). But the other notion we might expect, the *élan vital*, does appear as the source of moral progress – progress that, like the evolution in *Creative Evolution*, does not imply a goal.

But there is a certain ambivalence about the *élan* here. The closed morality, as we saw, had its origin and impetus in a social instinct reinforced and directed by social pressure. It is motivated by an impulsive force. The open morality on the other hand is motivated by an attractive force. It is a morality not of pressure but of aspiration. A vital role is played in it by the 'hero' or charismatic figure whom we are moved to follow, but the aspiration which draws us in this direction is the same as that which culminated in the human species i.e. the *élan* (MR, 42). This, however, was originally introduced (in *Creative Evolution*) as a *vis a tergo* to be contrasted with the *vis a fronte* of finalism, albeit also with that other *vis a tergo*, mechanism. Here the *élan* has more of a finalistic look about it, and this has some bearing on the claim sometimes made that in *Morality and Religion* the *élan* has become teleological, aiming to produce not just the human species but the 'open society' (embodying the open morality) as a state of that species. We have just seen that progress does not imply a goal (MR, 39), and: 'We do not assert that nature has, strictly speaking, designed or foreseen anything whatever' (MR, 43), though we can harmlessly follow the biologists in speaking of intention when we are really referring to function. Deliberative teleology then is not envisaged at least in this part of *Morality and Religion* (though for a claim that teleology does not involve deliberation see Aristotle, *Physics* II 8). Formally perhaps Bergson could escape contradiction by saying that the *élan* itself is not purposive but its operation involves our acting purposefully in so far as we are attracted by and follow heroes.

The transition from one morality to the other, which the *élan* engineers, involves a leap, as the Zeno point tells us, and Bergson uses the image of a set of iron filings suddenly taking on a certain shape because an invisible hand has been thrust into them, an image he had used back in *Creative Evolution.* (See CE, 99 – 100, and MR, 41, 94, 95.) But the two moralities are not unconnected. The closed morality lends something of its imperative character to the open and receives from it 'a connotation less strictly social, more broadly human' (MR, 37), or 'something of its perfume' (MR, 38). The 'perfume' culminates in what Bergson calls 'joy', which includes pleasure but goes beyond it and which he says little about (MR, 39, 45, 274 – 5).

The features of the open morality can be summed up then as follows: it involves aspiration rather than impulsion, is based on feeling or (higher) emotion rather than reason, is supra-intellectual rather than infra-intellectual, is universal rather than partial in its sympathies, follows individual example rather than rules, and leads to joy rather than mere pleasure.

One question that remains is in what sense it is a morality. Bergson more than once refers to 'complete' morality (MR, 23, 79), and evidently regards the open variety as a culmination and completion of the closed, though much of what he says about the open morality and the 'heroes' that embody it is connected, as he intends it to be, with mysticism. To most people, as Bergson is aware, mysticism suggest quietism, a preoccupation with personal visions or experiences, which may unite the mystic with the absolute but don't have much effect on his ordinary life, unless to take him out of it. This is a feature in fact of many systems looking to individual experiences or perfection, apart from those involving strictly religious mysticism. Plato's Guardians only re-entered the cave out of a sense of duty, and Aristotle's sage only entered politics as an *éminence grise* who could be consulted on the side-lines but was not expected to dirty his hands in civic affairs; Aristotle's contemplative God was notoriously uninterested in the world which was kept going by admiration and imitation of him.

At first sight Bergson's hero could be similarly remote, provided that he had certain features which when imitated would constitute action of a certain type in the world. The hero himself need not act, since imitation need not, and indeed cannot, imitate every feature of the model. The hero could be, say, calm and unruffled in his contemplations while his imitators were calm and unruffled in practical affairs. Bergson, however, does not have recourse to this device. Mysticism for him has as its end contact and so partial coincidence with the *élan* itself (MR, 188); it does not imply withdrawal but rather phrenetic activity. He has in mind religious leaders like St. Paul or Joan of Arc (MR, 194).

But what is this activity in aid of, and how is it related to morality? Most Christian mystics of this kind have sought to spread Christianity, though Bergson points to Joan of Arc as an exception. But what we are looking for is not how certain ideas or practices spread but what ideas they are. The impossibility, because of the Zeno point, of passing by simple expansion from group-loyalty to universalism suggests that universalism is indeed what we are looking for, or part of it. He speaks of the mystics as having 'love' for mankind, which suggests a concern for their interests, but he shows little concern with what the mystics conceive those interests to consist in. Some passages suggest joy is the answer: 'Pleasure and well-being are something, joy is more. . . . They mean, indeed a halt or marking time, while joy is a step forward' (MR, 45). But where to? Riches are an evil, says the Gospel; so why injure the poor by giving them our riches? Not for their sake, Bergson replies, but for our own. 'The beauty lies, not in being deprived, not even in depriving oneself, but in not feeling the deprivation' (MR, 45 – 6). But surely we ought to destroy our riches in that case, not inflict them on the poor? The question constantly arises: does the open morality consist in the universal application of principles, or pursuit of ends, previously applied or pursued only in a limited context, or are new principles to be applied and new ends pursued?

5. *Justice and value*

We have seen already that the two moralities are connected, and obligation survives into the second. Perhaps the nearest Bergson comes to an explicit discussion of what the connexion is appears in the eleven-page paragraph on justice (MR, 54 – 65). Here absolute justice is distinguished from relative, and the substance of justice from its form. Absolute justice is universal as against being limited to a peer-group, but this quantitative extension is based on a qualitative change, for the rights of the person as such are now regarded as inviolable, and not based on mercantile considerations (MR, 56 – 7). On the substance of justice we learn nothing directly except that it was the universalisation leap that was decisive for its progress (MR, 61), but the development of form evidently consisted in substituting justice as an end in itself for justice as merely something that met a social need and owed its status as obligatory to social pressure (MR, 60).

Here then a reasonably coherent picture seems to emerge. Justice starts from a social instinct, modified by the intellect in its applications, reinforced by social pressure, and limited to the area of the need. The *élan* responsible for this process then goes on to produce religious or mystical ideas which involve a 'love' for humanity that in turn produces a new motivation for justice, respect for mankind as such, thus universalising justice in a way that could not be achieved by mere expansion of the in-group so long as this relied on the foil of an out-group.

So the answer to the question ending our last section is that the same principles are to be applied but from a different motivation and on a different and wider basis.

Further questions remain, however. Justice cannot be an ultimate value, since it concerns assignment and distribution, so there must be something to be assigned and distributed. At the instinctive level from which Bergson's account begins it is presumably what promotes survival, especially of the species, that counts, and it is fair to assume that in general, though not necessarily in detail, evolution has connected this with pleasure: animals, even the self-sacrificing ants Bergson is so fond of, in general pursue what they find pleasant, though the social instinct may prompt them to do presumably unpleasant things. It seems reasonable to assume then that in the sort of system Bergson is constructing what justice is concerned with is in the first place 'the good things of life' as people in general would understand them.

It is unlikely that this would satisfy Bergson for long, but the question we are raising is one he scarcely tackles except in passing and by implication. It is all very well to talk of 'exceptional souls' arising who are 'borne on a great surge of love towards humanity in general' (MR, 77 – 8), while mystics are inspired by love to spread 'what they have received', where this love is 'an entirely new emotion, capable of transposing human life into another tone' (MR, 81). But what is to result from loving people with this sort of love? Not just that one gives them 'the good things of life'? The mystics want to spread 'what they have received', which is presumably either love itself, taken as having value in itself, or some sort of mystical experience. If Bergson is to be consistent he had better take this last line, and perhaps in the last resort this is what he intends. Love may have its value, but if it alone has value it is unclear how we could express it, and if other things have value but only love is to guide our actions we shall have the paradox of universal altruism: we shall give all our goods to the poor, who, observing that we are now poor, will promptly give them back again. But, if the mystical experience brings something called 'joy', conceived, as valuable, and love is simply a side-effect of this, then the love can consistently express itself in promoting joy, whether or not it also promotes other goods too, of a more utilitarian nature.

Earlier, towards the end of the long paragraph on justice,

Bergson mentions liberty and equality as what justice aims at. But he adds that these provide no clear guidance for action, because the extent to which a liberty will encroach on the liberty of others may not be ascertainable until we take account of the new ethos that this liberty itself will generate, while equality has to be balanced against liberty and so cannot provide an independent target (MR, 63 – 4). He does not expand on these points, but he could in any case add that liberty and equality presuppose other values, if they are to be of value themselves.

6.The hero

Finally in the sphere of ethics a word about the hero. Since the open morality depends on feeling rather than intellect, aspiration rather than pressure, it is perhaps not surprising that a model or embodiment of it has a large part to play, though necessarily a secondary part in that the initial hero could not himself arise by aspiration to such a model. If aspiration comes in at all in his case it must be towards embodying an abstract model. He is evidently a product of the *élan* and not to be further explained.

There is a problem about why when he does arise he has the charismatic effect he does. Mystics are usually regarded as unintelligible by most people, who would probably echo Dr Johnson's remark that if they have seen the unutterable they should not try to utter it. But Bergson casts his net rather widely when gathering mystics. He includes Joan of Arc, and St Paul, who would not be thought of as a mystic in the ordinary sense, despite his one special experience: it does not seem to have involved any union with the absolute. Bergson is evidently thinking of charismatic spiritual leaders in general. As for why they have the effect they do, he would presumably say this is because the *élan* which produces them also produces many near-misses, who are not quite spiritual leaders themselves but enough in tune with the model in question to be able to recognise it when it appears and to be triggered by it. As for how far the actions of such leaders can really be brought under the heading of 'love' rather than fanaticism, this is something Bergson says little about (how would he have reacted to Khomeini, for instance?), and perhaps we need spend no more time on it.

The mention of heroes suggests comparison with modern discussions of 'saints and heroes' and supererogation. There are differences, however. Bergson's hero does not sacrifice his own interests far beyond the call of duty, or do his duty in extreme circumstances, like the heroes and saints respectively of J.O. Urmson

(1958). No doubt the hero would sacrifice himself if necessary, like the ant to which he is distantly related. But the emphasis is not on a contrast between his actions and his self-interest but rather on the transfiguration of values that he embodies. Urmson's saints and heroes have as big a role to play in group-loyalty situations as anywhere else, but they would not there be Bergsonian heroes. Nor do Urmsonian saints and heroes need mystical experiences, and there is no emphasis on their transforming society in their role as exemplars.

7. Myth-making

The three distinctions, between closed and open morality, closed and open society, and static and dynamic religion, go essentially together. The closed society practises closed morality with a static religion; the open society practises (or would practise) open morality with a dynamic religion. The term 'dynamic' seems to refer to the effect of mysticism in transforming morality, while 'static' refers to the absence of this effect. Static religion is connected with morality, and the 'amoralism' of primitive religion results from comparing primitive religion with advanced morality; primitive morality is simply custom, and so coextensive with primitive religion (MR, 102). But religion and morality need not thenceforth advance hand in hand, and religion, while cementing the morality needed for social cohesion, can be carried in the direction of the particular kind of national development that a given society undergoes (MR, 175 – 6). However, though this may explain the harnessing of religion to nationalism ('with God on our side . . . '), it hardly explains the moral foibles of Zeus or even of Jehovah.

As obligation was needed to maintain social cohesion when instinct gave way to intelligence, so static religion, and in particular the myth-making element in it, grew up as a counterweight to the subversive tendencies of intelligence, which left to themselves would have led to egoism and social disruption. We reach this conclusion by asking what function myth-making plays, for as we determine function by looking at structure, so also we can only determine structure by looking at function; otherwise we shall not know how to organise our data (MR, 88, 97). Is this an example of finalism? If so, it is of a kind common enough among scientists. It is also only 'as if' nature has intentions (MR, 90). (Cf. MR, 96, 270.) But function will only explain the survival of structure, not its origin, so how does myth-making originate? Here Bergson is less clear, and though he explicitly raises the question of origins he ends up by simply appealing to the *élan* (MR, 90 – 1).

Myth-making arises as a reaction to intelligence, and yet at a time when instinct is being superseded, so how is it related to them? Bergson appeals again to the notion of 'virtual instinct' apparently meaning that it shows intelligence producing results like those instinct produces in insect societies (MR, 91).

But, if myth-making is the work of intelligence, why are its results so at variance with what our modern intelligence seems to tell us? Bergson's treatment of this leads him to some of the most interesting discussions in the book. Should one follow Lévy-Bruhl, a friend of his youth (MR, 127), in seeing a special primitive logic or mentality at work? Bergson thinks not. For one thing, we think in superstitious ways ourselves, as we can see by observing our own reactions in the roulette saloon (MR, 117 – 18); it is greater knowledge, not a different mentality, that inhibits our tendency to superstition for most of the time. On the other hand primitives are perfectly well able to cope with the world around them at the practical level. 'Mystic' or occult causes are never appealed to when dealing with the inanimate as such, but only when human interests are at stake. Chance is itself a concept we only use in this sort of case, and when Lévy-Bruhl accuses primitives of not believing in chance Bergson turns the tables on him by insisting that to believe in chance is to exhibit the kind of outlook he distinguishes as primitive (MR, 123). If a tile falls on a man and kills him, we may know why it fell, and even why it fell then (we may add), provided 'then' is specified neutrally, e.g. by clock-time. But why did it fall *when the man was passing*? This is what suggests intention, and this suggestion remains when we attribute it to chance, for we would not bring in chance to explain its falling at (say) 12.43, unless that moment was significant for us. 'Chance is . . . an intention emptied of its content' (MR, 124). (Cf. Sorabji 1980: chapter 1, discussing Aristotle on coincidence.) Bergson refers to his own 1900 – 1 lectures on Alexander's *De Fato*. ('1898' in his footnote is a slip; see Q, 1571, M, 438.)

So far we have seen Bergson arguing, with some plausibility, that 'primitive' attitudes to natural events have their echo in all of us. But what is a mere inchoate tendency in modern man may be a gross and explicit superstition taken fully seriously by pre-scientific man. How shall we explain these superstitions on the part of people no less intelligent than ourselves? Bergson does not define superstition, and it is tempting to treat it as simply the other chap's religion. More seriously we might try defining a belief as superstitious if it contradicts, rather than merely goes beyond, the known results of science, including its negative results (that there are no fairies and no correlations between misfortune and walking under ladders on Friday the thirteenth). We could allow miracles to escape if we add that the belief should not belong to a frame of thought that predicts systematic and explicable breaches of natural laws. Bergson's explanation again appeals to function. Superstition, and myth-making in general, exists to counteract certain unwanted side-effects of the growth of intelligence, notably egoism, and again the origin of it is attributed to 'the residue of instinct which survives on the fringe of intelligence' (MR, 99). The function of discouraging egoism is served by the invention of gods as a moral sanction, an ancient idea that goes back at least to Critias in the late fifth century B.C., though Bergson avoids the crudity of having the rulers consciously invent the gods. On the individual level myth-making counters the fear of death by assuring an afterlife – Bergson hastens to add that only in the primitive form is such assurance illusory (MR, 109 n.). Of the possibility of an afterlife men became convinced, he thinks, by dissociating the visual image of the body, as seen in reflections etc., from the tactile image (MR, 110 – 11). Apart perhaps from this last point the content of these views is hardly original, but the frame in which he puts them is Bergsonian. Intelligence must be kept from going too far by being presented with counterfeit facts, for, as he states with considerable exaggeration, 'the finest arguments in the world come to grief in the face of a single experiment: nothing can resist facts', and: 'A fiction, if its image is vivid and insistent, may indeed masquerade as perception and in that way prevent or modify action' (MR, 89). Since a false experience can stop intelligence drawing too many conclusions from its true experiences, 'we should not be surprised to find that intelligence was pervaded, as soon as formed, by superstition' (ibid.). Again appeal to function is sufficient as an explanation, and the fiction is simply attributed to the 'residue of instinct' (MR, 99). That religion has often served as a convenient 'opium of the people' and that the belief in survival owes much to wishful thinking we need not deny. But it is hard to see anything like a sufficient explanation here for the extraordinary elaborations that myths have undergone, though Bergson might (and presumably must) allow clashes between the 'social control' and 'reassurance' motifs if he is to explain such things as the extraordinarily unpleasant nature of many visions of the afterlife, and he does allow that genuinely primitive religion was very different from anything we find now; he constantly emphasises that contemporary 'primitives' have as long a tradition behind them as we have.

8. Magic

Another element in human thinking which Bergson seeks to explain is magic. His approach is that of armchair anthropology, with little attempt to consider empirical data in any detail – one may contrast his attitude to aphasia which we mentioned in chapter 5. Considered within this limitation it is a strength of his approach that he tries to derive magic from the emotional and active sides of our nature rather than from some strange distortion of our intellect; he says of the principle that 'like begets like': 'There is no apparent reason why humanity should begin by positing so abstract and arbitrary a law' (MR, 142). Magic begins when intelligence 'translates into a conception' what instinct suggests (MR, 140). Under the influence of rage we carry out movements as of strangling our enemy, but realising that the result is incomplete we *want* the material world to complete the result, which it will do if invested with powers that yield to human desires. We therefore attribute to matter the sort of nature expressed by words like 'mana' or 'wakonda'. Magic is thus 'the outward projection of a desire' (MR, 141).

Without waiting to become so enraged that we go through the motions of strangling someone we can see the plausibility of this approach by considering how we instinctively press our feet down in a car being driven too fast, or 'will' our golfball to enter the hole it is uncertainly approaching; Bergson insists that magic is not something confined to a bygone age (MR, 146). The weakness of the account comes when we inspect more closely the attributing to nature of such convenient powers. He insists that the practice of magic precedes rather than follows such conceptions as that of mana; it does not depend on a pre-existing intellectual 'pre-logic' (MR, 145 – 6). But he does not make clear how the practice gets going without this. Perhaps the concept of mana as an explicit general concept only arises from reflection on already existing magical practices, but the step from wanting nature to complete our stranglings for us to acting as though it does and believing that it does is still unexplained. 'The essential [in magic] is always to repeat in tranquillity, with the conviction that it is efficacious, the act which has given a quasi-hallucinatory impression of its efficacy when performed in a moment of excitement' (MR, 143). Rather than by an effort of imagination raising himself to the clouds the rain-maker will find it 'simpler' to pour out a little water (ibid.). But simplicity alone will hardly be enough.

What Bergson does emphasise, however, is that magic is neither proto-science nor proto-religion. Magic, like science, has had to grow, and is not a forerunner of science but a lazy substitute for it. Similarly magic and religion, though not entirely disconnected, 'go their separate ways, having started from a common origin', presumably the desire to influence nature (MR, 148). Religion personifies what man appeals to while magic depersonifies it, and neither is the forerunner of the other.

9. Mysticism

Static religion defends humanity against the unwanted side-effects of intelligence, the apprehension which rational foreknowledge produces and which leads in turn, so Bergson claims, to 'a momentary slackening of the attachment to life'. But if static religion 'attaches man to life, and consequently the individual to society', man could recover his confidence instead 'by turning back for fresh impetus, in the direction whence that impetus came', by fastening on the 'fringe of intuition' that surrounds intelligence and consummating it in action (MR, 179 – 80). In other words man could appeal instead to dynamic religion culminating in mysticism.

Bergson then embarks on a long discussion of the historical development of mysticism, of its relations to static religion, and of the justification for using 'religion' of both the static and dynamic varieties. Religion is a sort of crystallisation of mysticism and stands to it as popularisation stands to science (MR, 203 – 4).

We have already seen something of the nature and role of mysticism, but what of its status? Is it more than an illusion? This will depend on what claims it makes. In earlier works Bergson said little about God, as we saw in chapter 7. But in later life he became a Catholic, whether or not, out of solidarity with his persecuted fellow Jews, he refrained from formal conversion: this is disputed. In *Morality and Religion* then God becomes more prominent, albeit virtually confined to the last half of the third chapter, and mystical experience is treated as an argument for His existence.

Bergson realises that two questions arise: Is mystical experience evidence for anything? And is what it is evidence for what we normally mean by 'God'? The second question depends on what we do mean by God, and it is not surprising that Bergson thinks that the God revealed by mysticism is very much nearer to the living and communicating God of ordinary thought than is the abstract and changeless God of the philosophers, of whom he takes Aristotle as a prime example. The God of mysticism, at least as conceived by Bergson, is far more fitted than Aristotle's self-thinking thought to be a God for a philosophy of change. One might wonder what Plotinus would say about this. But Plotinus, though he could look upon the promised land, could never set foot upon its soil. He reached contemplative ecstasy, but not the stage where 'the human will becomes one with the divine will' (MR, 188). However, similarity is not identity. That Bergson's God has in common with the God of popular thought activity (in the ordinary sense, not Aristotle's *energeia*) does not imply He is a personal God in the required sense. In fact a certain paradox arises here akin to what happens when we treat naive realism as a philosophical view. Bergson may be a philosopher of change, but he is a philosopher: he is not writing on a Clapham omnibus. How then can he, any more than any other philosopher, do justice to an essentially non-philosophical God?

We shall see more in a minute about Bergson's views on the nature of God, but let us first turn to the other question: is mystical experience evidence for anything? Bergson first considers the objection that mystics are of unsound mind. He replies that they are people of action and sound common sense, that what seem to be symptoms of unsound mind may be due simply to their being under great stress, that we do not deny genius its due because it is often accompanied by eccentricities, that mystics should not be blamed for the antics of pseudo-mystics, for anything can be parodied (MR, 194 – 6, 209 – 10). These points no doubt have some force, though they cannot go far until supplemented by positive arguments, and they hover between saying mystics don't behave oddly and that they do but for extraneous reasons; one could add that a fanatical *idée fixe* is quite compatible with success in practical affairs.

Another objection is that mystical claims are not publicly verifiable (MR, 210 – 11). Bergson replies that at one time claims about Central Africa were only verifiable by intrepid explorers. He admits that in principle others could go too, but so, he thinks, they could with mysticism, and that most of us could get at least a short way. No doubt most of us could get a few miles behind the sea-ports of Africa, but there is no sign yet of any satellite mapping of the land of the mystics. What Bergson ignores here is that no-one denies that mystics have experiences of the kind they claim to have. The question is, are those experiences evidence of a reality behind them? Suppose Central Africa could *only* be described in terms of secondary qualities: what status would we give to claims about it? Satellites wouldn't help. Bergson incautiously takes music as an example: we don't take the existence of the tone-deaf as an 'argument against music'. But what would an 'argument against music' be, or an argument for it, for that matter? What the tone-deaf should not deny is that musical experiences exist, and can be of great value to those that have them. Let us allow the same about mystical experiences; but what follows?

Bergson next appeals to the agreement that exists among mystics (MR, 211). He admits that this might be explained as due to common influence, but thinks this would only explain external resemblances; he has indeed himself spent many pages describing different kinds of mysticism in different religions etc., but he claims they are all stages on a single road. If many mystics all claim to hold intercourse with a certain Being, the existence of that Being is the simplest hypothesis to explain this (MR, 212).

How powerful is this 'inference to the best explanation'? The blind are well advised to rely on agreement among the sighted, though the blind may have no conception of how this agreement comes about; the sighted can make predictions which the blind can verify without acquiring sight. Mystics may predict each other's experiences, and we may think their success rate cannot be explained by mere guessing. But the hypothesis that a certain Being exists will only explain things if it is itself coherent. If the hypothesis that there is a God is to be genuinely explanatory God must presumably be more than a logical construction; He must have objective independent existence. But what sense can we non-mystics make of a God who can only be described in language unintelligible to us in a way colours are to a blind man?

But perhaps we are going too far. Colours are not entirely unintelligible to the blind, who have associated concepts like shape and can understand accounts of the mechanism of sight. Is not the non-mystic even better placed, since mystics do not usually use words (like colour-words) that non-mystics cannot understand; they speak in English or whatever. But words are not the only bearers of meaning; their combination must be taken into account as well. 'Colourless green ideas sleep furiously' (Chomsky) uses only common English words. As for the appeal to associated concepts, we must ask for examples. Suppose the mystic predicts that if I pray in a certain way (let us assume this is unproblematic) I shall get an experience of a kind I do understand, e.g. of being comforted, or of uplift, and suppose this happens. Then, says the mystic, God must exist as the person who comforts or uplifts. But what concept of a person have I except of something that acts in the physical world, and predictions of such divine intervention (saving the innocent, punishing the guilty, etc.) are notoriously shaky. How can I distinguish experiencing an objective God from having some sort of illusion? Compare the question how a disembodied spirit could distinguish perceiving from imagining.

But is a physical world really necessary for objectivity? Could not mathematicians agree on objective truths even of a kind that had no physical application? Yes, but they do so by applying laws of logic with general application to terms that initially raise no (relevant) problems of intelligibility. It is a mere contingency that we lack the intelligence to follow them all the way.

10.God

Bergson may have reached, or returned to, God in his old age, but the God he reached does not seem to have been all that orthodox. He insists that God *is* love, not just loving or lovable, but without, I think, throwing any more light on this murky equation than most writers do. This love has an object – us – because 'God needs us, just as we need God' and He created creators so as to have beings worthy of His love (MR, 218). Not only does God have needs, but even His almightiness seems in doubt, perhaps because of the not very clear relations between Himself and the *élan*. The *élan* itself is certainly not almighty. It is finite (MR, 44) and Bergson constantly talks of it as meeting obstacles which it circumvents with ingenuity and inventiveness. At one point a 'great current of creative energy is precipitated into matter, to wrest from it what it can' (MR, 178), which suggests matter pre-exists the *élan*, though 'matter and life, as we define them, are coexistent and interdependent' (MR, 219). Back in 1908 Bergson had insisted that God is distinct from and is the source of the *élans* which form worlds, and that only God has always existed (so that the senselessness of supposing that there might have been nothing does not show that anything except God must exist) (M, 766 – 7). Now, however, God is said to *be* the creative energy which brings the universe into being, while men 'had to be' wrought into a species, involving other species too (MR, 220 – 1).

Bergson faces the omnipotence question explicitly after a refreshingly realistic discussion of the problem of evil (MR, 223 – 5). Resolutely rejecting all Panglossian solutions he sees two grounds for optimism, that men must find life good on the whole since they cling to it, and that mystics can attain a state of bliss. The first point is fallacious. That men do not commit suicide does not imply that they find life worth living in the sense of being above the point of indifference, for they could be inhibited by a purely instinctive fear of death which only horrors far below the neutral point could overcome. (Bergson is not alone in this fallacy. It vitiates much of Parfit 1984.) But he does not rest content with these grounds for optimism. Should not an omnipotent God have prevented all suffering? No, for if the notion of 'everything' implicit in 'omnipotent' refers to all possibilities it is as senseless as the notion of 'nothing'. Bergson does not expand this point, whose relevance seems to be that God only had a *choice* about which possible world to actualise if the possible worlds can be laid out as a definite set of alternatives, which Bergson thinks they cannot be. One might perhaps compare G. Schlesinger's argument that God cannot be blamed for not creating the best possible world because there is no such thing. (Schlesinger 1977: part I).

11.Epilogue

In a book on Bergson's philosophy I must pass over the final chapter of *Morality and Religion* with its at first sight unpromising suggestion of mysticism as a remedy for the discontents of industrial society. Philosophical material crops up here and there, but the main point is that industrialisation has led not just to uniformity – a small price to pay if it liberated us from drudgery – but to artificial wants, which become treated as necessary and lead to wars. The remedy lies in a return to a simpler life, and the chapter contains some hints of the active part Bergson took in promoting education and mutual understanding between nations as the best prophylactic against war. It is ironic and sad that this contrasts so markedly with his jingoistic attitude in both the world wars. But it is intriguing to wonder, in view of his call for a return to the simpler life, whether today his vote would go to the Greens.

Bibliography

The first section of what follows lists the editions and translations I have used of Bergson's main works, together with the abbreviations I have given them (listed separately at the beginning of the present book). Except for *The Creative Mind* and *Duration and Simultaneity* all the translations were made with the approval of Bergson himself. The second section lists works referred to, together with one or two others of interest or which I am indebted to. For a full bibliography see Gunter 1986.

Works by Bergson.

Oeuvres (Q), Paris, Presses Universitaires de France, 1959. (This 'Edition du Centenaire' was re-issued with additional indexes in 1970. It is edited by A. Robinet with an 'Introduction' by H. Gouhier. It contains the following works, listed here with date of first publication and English translation, and its margins have page references to the standard editions of Bergson's works (see Q, 1484).)

Essai sur les données immédiates de la conscience, 1889 (*Time and Free Will*).
Matière et mémoire, 1896 (*Matter and Memory*).
Le Rire, 1900 (*Laughter*).
L'Evolution créatrice, 1907 (*Creative Evolution*).
L'Energie spirituelle, 1919 (*Mind-Energy*).
Les Deux Sources de la morale et de la religion, 1932 (*The Two Sources of Morality and Religion*).
La Pensée et le mouvant, 1934 (*The Creative Mind*). (This includes *inter alia* 'Introduction à la métaphysique', first published in 1903.)
Mélanges (M), Paris, Presses Universitaires de France, 1972. (Edited by A. Robinet with an 'Avant-Propos' by H. Gouhier, this volume contains Bergson's early thesis *L'Idée de lieu chez Aristote*, 1889, the book *Durée et simultanéité*, 1922 (*Duration and Simultaneity*), and virtually all the writings whose publication Bergson did not veto in his

will. These, arranged in chronological order, include articles, preliminary versions of parts of his books, reviews, lectures and addresses, and letters. They cover a wide variety of topics as well as philosophy. All translations from *Mélanges* are my own unless otherwise stated.)

Time and Free Will (TF), transl. F.L. Pogson, London, Allen & Unwin, 1910.

Matter and Memory (MM), transl. N.M. Paul and W.S. Palmer, London, Allen & Unwin, 1911.

Laughter (L), transl. C. Brereton and F. Rothwell, London, Macmillan, 1911.

Creative Evolution (CE), transl. A. Mitchell, London, Macmillan, 1911.

Mind-Energy (ME), transl. H.W. Carr, London, Macmillan, 1920. (Lectures and essays mainly on the results of Bergson's researches into problems of mind and body.)

Duration and Simultaneity, transl. L. Jacobson, with an introduction by H. Dingle, Indianapolis, Bobbs-Merill (Library of Liberal Arts), 1965. (This translation is apparently uncommon in the UK, perhaps because of doubts about the original (see above, chapter 2 §13). I have referred to the French original in *Mélanges*.)

The Two Sources of Morality and Religion (MR), transl. R.A. Audra and C. Brereton with the assistance of W.H. Carter, London, Macmillan, 1935.

The Creative Mind (CM), transl. M.L. Andison, New York, Philosophical Library, 1946. (Articles and lectures, officially on philosophical method. The volume contains a long double 'Introduction' specially written for it, and also 'Introduction to metaphysics', previously translated by T.E. Hulme (New York, Putnam's Sons, 1912); see also Gunter 1986: 36. My own references are to the current volume. The paperback edition has different paging and omits the footnotes, with no indication of either of these facts. As it is quite common in the UK I have given the hardcover references followed in brackets by the paperback references or, in the case of the footnotes, the references to the French text in *Oeuvres*. This book is perhaps the best starting-point for the newcomer to Bergson.)

Other references

Writers before 1800 are included only where a specific edition is referred to. I have not attempted to cover all reprintings.

Augustine (1964), *Confessions* XI 15 in Smart, J.J.C. (ed.), *Problems of Space and Time,* New York, Macmillan, London, Collier-Macmillan. (For an earlier translation see Gale, R.M. (ed.), *The Philosophy of Time*, London, Macmillan, 1968. Cf. also Findlay, J.N., 'Time: a treatment of some puzzles', *Australasian Journal of Psychology and Philosophy,* vol.XIX, no.3, pp.216–35, reprinted in both Smart and Gale.)

Barlow, M. (1966), *Henri Bergson,* Paris, Presses Universitaires de

France. (Brief French biography which integrates life with works quite well.)

Barreau, H. (1973), 'Bergson et Einstein: à propos de *Durée et simultanéité*', *Les études Bergsoniennes*, vol.X; pp.73-134.

Berthelot, R. (1913), *Un romantisme utilitaire: étude sur le mouvement pragmatiste. Troisième partie. Un pragmatisme psychologique: le pragmatisme partial de Bergson*, Paris, Alcan.

Bjelland, A.G. (1974), 'Bergson's dualism in *Time and Free Will*', *Process Studies*, vol.IV, no.2, pp.83–106. (Soft-pedals it.)

Boudot, M. (1980), 'L'espace selon Bergson', *Revue de Métaphysique et de Morale*, vol.85, no.3, pp.332–56.

Broad, C.D. (1938), *Examination of McTaggart's Philosophy*, Cambridge, Cambridge University Press. (Vol.2, pp.281–8, discusses specious present.)

Brooks, D.R. and Wiley, E.O. (1984), 'Evolution as an entropic phenomenon', in Pollard, J.W. (ed.), *Evolutionary Theory: Paths into the Future*, Chichester, John Wiley & Sons.

Brooks, D.R. and Wiley, E.O. (1986), *Evolution as Entropy*, Chicago and London, University of Chicago Press.

Buchdahl, G. (1961), 'The problem of negation', *Philosophy and Phenomenological Research*, vol.XXII, no.2, pp.163–78.

Čapek, M. (1969), 'Bergson's theory of matter and modern physics', in Gunter, P.A.Y. (ed. and transl.), *Bergson and the Evolution of Physics*, Knoxville, University of Tennessee Press.

Čapek, M. (1971), *Bergson and Modern Physics*, Dordrecht, Reidel.

Čapek, M. (1980), 'Ce qui est vivant et ce qui est mort dans la critique bergsonienne de la relativité', *Revue de Synthèse*, vol.CI, nos.99-100, pp.313-44.

Carr, H.W. (1914), reply to Russell, included in Russell, B., *The Philosophy of Bergson*, Cambridge, Bowes & Bowes. (Originally published elsewhere in 1912.)

Chambers, C.J. (1974), 'Zeno of Elea and Bergson's neglected thesis', *Journal of the History of Philosophy*, vol.XII, no.1, pp.63–76. (Discusses Bergson's early thesis on place in Aristotle, throwing some light on his development and also discussing his relations with Leibniz and Kant.)

Chevalier, J. (1928), *Bergson*, transl. L.A. Clare, London, Rider. (French original, 1926. Exposition of Bergson's philosophy which had his warm approval.)

Chevalier, J. (1959), *Entretiens avec Bergson*, Paris, Plon. (Record of conversations over many years by Bergson's Boswell. Useful as a source, though it tends to concentrate on Chevalier's own interests.)

Chisholm, R.M. (1966), 'Freedom and action', in Lehrer, K. (ed.), *Freedom and Determinism*, New York, Random House.

Chopra, Y. (1963), 'Professor Urmson on "Saints and heroes"', *Philosophy*, vol.XXXVIII, no.144, pp.160-6.

Cunningham, G.W. (1916), *A Study in the Philosophy of Bergson*, New York and London, Longman, Green.

Davidson, D. (1967), 'Causal relations', *Journal of Philosophy*, vol.LXIV, no.21, pp.691-703. (Often reprinted.)

Dawkins, R. (1986), *The Blind Watchmaker*, Harlow, Longman. (Defends Darwinism.)

Dennett, D. (1984), *Elbow Room*, Oxford, Clarendon Press.

Dobzhansky, T. (1955), *Evolution, Genetics, and Man*, London, Wiley.

Eldredge, N. (1986), *Time Frames*, London, Heinemann.

Findlay, J.N. (1941), 'Time: a treatment of some puzzles', *Australasian Journal of Psychology and Philosophy*, vol.XIX, no.3, pp.216–35. (Reprinted in Smart, J.J.C. (ed.), *Problems of Space and Time*, New York, Macmillan, London, Collier-Macmillan, 1964, and in Gale, R.M. (ed.), *The Philosophy of Time*, London, Macmillan, 1968.)

Frege, G. (1952), 'Negation', in Geach, P. and Black, M. (eds), *Translations from the Philosophical Writings of Gottlob Frege*, Oxford, Blackwell. (Original 1919.)

Fuss, P. (1963/4), 'Conscience', *Ethics*, vol.LXXIV, no.2, pp.111-20.

Gale, R.M. (ed.), (1968), *The Philosophy of Time*, London, Macmillan.

Gale, R.M. (1973-4), 'Bergson's analysis of the concept of nothing', *The Modern Schoolman*, vol.LI, pp.269-300.

Ginnane, W.J. (1960), 'Thoughts', *Mind*, vol.LXIX, no.275, pp.372–90.

Goodman, N. (1954), *Fact, Fiction and Forecast*, London, Athlone. (Revised in later editions.)

Goudge, T.A. (1967), 'Bergson' in Edwards, P. (ed.), *The Encyclopedia of Philosophy*, vol.1, New York, Macmillan, London, Collier-Macmillan.

Gouhier, H. (1959), 'Introduction', in Bergson, H., *Oeuvres*, Paris, Presses Universitaires de France, 1959.

Green, T.H. (1883), *Prolegonema to Ethics*, Oxford, Clarendon Press.

Gunter, P.A.Y. (ed. and transl.) (1969), *Bergson and the Evolution of Physics*, Knoxville, University of Tennessee Press.

Gunter, P.A.Y. (1986), *A Bibliography of Bergson*, revised second edition, Bowling Green, Philosophy Documentation Centre. (First edition 1974.)

Gunter, P.A.Y. (1987): See Papanicolaou, A.C. and Gunter, P.A.Y.

Hamlyn, D.W. (1961), *Sensation and Perception*, London, Routledge and Kegan Paul. (See p.170.)

Heidsieck, F. (1957), *Henri Bergson et la notion d'espace*, Paris, Le Circle du Livre. (Influential in the recent rehabilitation of Bergson. See especially chapter 7 on *Durée et simultanéité*.)

Hobart, R.E. (1934), 'Free will as involving determination and inconceivable without it', *Mind*, vol.XLIII, no.169, pp.1 – 27.

Holland, R.F. (1954), 'The empiricist theory of memory', *Mind*, vol.LXIII, no.252, pp.464 – 86.

Hume, D. (1888), *A Treatise of Human Nature*, ed. by L.A. Selby-Bigge, Oxford, Clarendon Press. (Originally published 1739-40. For 'distinctions of reason' see 1.1.7.)

Husson, L. (1947), *L'Intellectualisme de Bergson*, Paris, Presses Universitaires de France.

James, W. (1890), *The Principles of Psychology*, London, Macmillan.

James, W. (1909), *A Pluralistic Universe*, London, Longmans, Green.

James, W. (1912), *Essays in Radical Empiricism*, London, Longmans, Green. (For neutral monism see chapter 1, 'Does "consciousness" exist?', originally published in 1904.)

Jankélévitch, V. (1959), *Henri Bergson*, Paris, Presses Universitaires de France.

Janvier, P. (1984), 'Cladistics: theory, purpose and evolutionary implications', in Pollard, J.W. (ed.), *Evolutionary Theory: Paths into the Future*, Chichester, John Wiley and Sons.

Kant, I. (1873), *Critique of Practical Reason*, transl. T.K. Abbott, London, Longmans, Green. (Original edition, 1788. For the discussion of the deposit Bergson's editor refers to 1.1 §4, Theorem III, Remark (Q, 1570); see p.115 in Abbott.)

Kirk, G.S. and Raven, J.E. (1957), *The Presocratic Philosophers*, Cambridge, Cambridge University Press. (In the second edition of this work (1983) the chapter on Zeno is rewritten by M. Schofield.)

Kolakowski, L. (1985), *Bergson*, Oxford, Oxford University Press. (Useful brief introduction viewing wood rather than trees.)

Lacey, H.M. (1970), 'The scientific intelligibility of absolute space', *British Journal for the Philosophy of Science*, vol.21, pp.317 – 42.

Le Dantec, F. (1907), 'La biologie de M. Bergson', *La Revue du Mois*, vol.IV, no.20 (August), pp.230 – 41. (Bergson replies in vol.IV, no.21 (September), pp.351 – 3, also in Bergson, H., *Mélanges*, Paris, Presses Universitaires de France, 1972, 731.)

Lenneberg, E.H. (1967), *Biological Foundations of Language*, New York, Wiley.

Lindsay, A.D. (1911), *The Philosophy of Bergson*, London, J.M. Dent & Sons.

Locke, D. (1971), *Memory*, London, Macmillan.

MacKay, D.M. (1958), 'Complementarity', *Proceedings of the Aristotelian Society*, supplementary vol.XXXII, pp.105 – 22.

McTaggart, J.M.E. (1908), 'The unreality of time', *Mind*, vol.XVII, no.68, pp.457 – 74.

MacWilliam, J. (1928), *Criticism of the Philosophy of Bergson*, Edinburgh, T. & T. Clark.

Martin, C.B. and Deutscher, M. (1966), 'Remembering', *Philosophical Review*, vol.LXXV, pp.161 – 96.

Milet, J. (1974), *Bergson et le calcul infinitésimal*, Paris, Presses Universitaires de France. (Cf. also J. Ullmo's preface, and review by J. Theau in *Dialogue*, vol.XV, no.1, 1976, pp.169 – 73).

Milet, J. (1987), 'Bergsonian epistemology and its origins in mathematical thought', in Papanicolaou, A.C. and Gunter, P.A.Y. (eds), *Bergson and Modern Thought*, Chur, Harwood Academic Publishers.

Moore, E.F. (1964), 'Mathematics in the biological sciences', *Scientific American*, vol.211, no.3, pp.148-64. (See pp.148, 156.)

Moore, G.E. (1903), *Principia Ethica*, Cambridge, Cambridge University Press. (See §§18 – 22 for organic unities.)

More, H. (1679): Henrici Mori, *Scriptorum Philosophicorum* Tomus Alter,

London, Martyne, J. and Kettilby, G. (The passage comes in More's second letter to Descartes, on p.248. Descartes replies on p.252.)

Mossé-Bastide, R.M. (1955), *Bergson Educateur*, Paris, Presses Universitaires de France.

Mouton, D.L. (1969), 'Thinking and time', *Mind*, vol.LXXVIII, no. 309, pp.60 – 76.

Newton, I. (1964), 'Absolute space and time', in Smart, J.J.C. (ed.), *Problems of Space and Time*, New York, Macmillan, London, Collier-Macmillan. (This is an extract from Newton's *Principia*, originally published in 1686. The title is Smart's.)

Papanicolaou, A.C. and Gunter, P.A.Y. (eds) (1987), *Bergson and Modern Thought*, Chur, Harwood Academic Publishers. (Essays whose main tenor is to emphasise the extent to which Bergson anticipated and finds confirmation in modern scientific theories, especially in physics and psychology and also parapsychology.)

Parfit, D. (1984), *Reasons and Persons*, Oxford, Oxford University Press.

Pollard, J.W. (ed.) (1984), *Evolutionary Theory: Paths into the Future*, Chichester, John Wiley and Sons.

Putnam, H. (1975 – 6), 'What is "realism"?' *Proceedings of the Aristotelian Society*, N.S. vol.LXXVI, pp.177 – 94.

Ridley, M. (1985), *The Problems of Evolution*, Oxford, Oxford University Press.

Robinet, A. (1965), *Bergson et les métamorphoses de la durée*, Paris, Editions Seghers.

Rohrlich, F. (1987), *From Paradox to Reality*, Cambridge, Cambridge University Press. (Very clear layman's introduction to the basics and significance of relativity theory and quantum physics.)

Ruse, M. (1986), *Taking Darwin Seriously*, Oxford, Blackwell.

Russell, B. (1914), *The Philosophy of Bergson*, Cambridge, Bowes & Bowes. (This volume also contains a reply by H.W. Carr and a rejoinder to it by Russell. These three items were originally published separately in 1912 and 1913. Russell's views are repeated in abbreviated form, with new initial paragraph, in his *History of Western Philosophy*, London, Allen & Unwin 1946. Bergson found Carr's reply 'excellente'.)

Ryle, G. (1929), 'Negation', *Proceedings of the Aristotelian Society*, supplementary vol.IX, pp.80 – 96.

Salmon, W.C. (ed.) (1970), *Zeno's Paradoxes*, Indianapolis and New York, Bobbs-Merrill. (Modern discussions with extensive bibliography.)

Schlesinger, G. (1977), *Religion and Scientific Method*, Dordrecht, Reidel.

Sciama, D. (1986), 'Time "paradoxes" in relativity', in Flood, R. and Lockwood, M. (eds), *The Nature of Time*, Oxford, Blackwell. (Discusses rocket paradox, calling it 'clock' paradox.)

Shoemaker, S. (1969), 'Time without change', *Journal of Philosophy*, vol.LVI, no.12, pp.363 – 81.

Simpson, G.G. (1949), *The Meaning of Evolution*, New Haven, Yale University Press.

Smart, J.J.C. (ed.) (1964), *Problems of Space and Time*, New York, Macmillan, London, Collier-Macmillan.

Sorabji, R.R.K. (1980), *Necessity, Cause and Blame*, London, Duckworth.

Steele, E.J., Gorczynski, R.M. and Pollard, J.W. (1984), 'The somatic selection of acquired characteristics', in Pollard, J.W. (ed.), *Evolutionary Theory: Paths into the Future*, Chichester, John Wiley & Sons.

Stewart, J.McK. (1911), *A Critical Exposition of Bergson's Philosophy*, London, Macmillan.

Strawson, P.F. (1959), *Individuals*, London, Methuen. (See chapter 2 for discussion of sounds. Cf. also Strawson (1980).)

Strawson, P.F. (1967), 'Is existence never a predicate?', *Critica*, vol.1, reprinted in his *Freedom and Resentment*, London, Methuen, 1974.

Strawson, P.F. (1980), 'Reply to Evans,' in van Straaten, Z. (ed.), *Philosophical Subjects*, Oxford, Clarendon Press, pp.273 – 82.

Taylor, R. (1955), 'Spatial and temporal analysis and the concept of identity', *Journal of Philosophy*, vol.LII, no.22, pp.599 – 612, reprinted in Smart, J.J.C. (ed.), *Problems of Space and Time*, New York, Macmillan, London, Collier-Macmillan.

Theau, J. (1976), Review of Milet, J., *Bergson et le calcul infinitésimal*, in *Dialogue*, vol.XV, no.1, pp.169 – 73.

Tye, M. (1986), 'The subjective qualities of experience', *Mind*, vol.XCV, no.377, pp.1 – 17.

Ullmo, J. (1974), 'Préface' to Milet, J., *Bergson et le calcul infinitésimal*, Paris, Presses Universitaires de France.

Urmson, J.O. (1958), 'Saints and heroes', in Melden, A.I. (ed.), *Essays in Moral Philosophy*, Seattle and London, University of Washington Press.

Waismann, F. (1949 – 53), 'Analytic-synthetic' (in six parts), *Analysis*, vol.10, no.2, pp.25 – 40, vol.11, no.2, pp.25 – 38, no.3, pp.49 – 61, no.6, pp.115 – 24, vol.13, no.1, pp.1 – 14, no.4, pp.73 – 89, reprinted in his *How I See Philosophy*, London, Macmillan, 1968. (See *Analysis*, vol.11, 1951, pp.50 – 2.)

Whorf, B. (1968), 'An American Indian model of the universe', in Gale, R.M. (ed.), *The Philosophy of Time*, London, Macmillan. (Originally published 1950.)

Winch, P.G. (1982), 'Ceasing to exist', *Proceedings of the British Academy*. (Reprinted in his *Trying to Make Sense*, Oxford, Blackwell, 1987.)

Wisdom, J. (1931), 'Logical constructions', *Mind*, vol.XL, no.158, pp.188 – 216, no.160, pp.460 – 75. (See p.193 for meanings of 'is a mammal' (without the example). Wisdom attributes the view to G.E. Moore – but see also Moore in Schilpp, P.A. (ed.), *The Philosophy of Bertrand Russell*, Chicago and Evanston, Northwestern University Press, 1944, pp.219 ff., esp. p.222.)

Wittgenstein, L. (1953), *Philosophical Investigations*, Oxford, Blackwell.

Zemach, E.M. (1979), 'Four ontologies', in Pelletier, F.J. (ed.), *Mass Terms: Some Philosophical Problems*, Dordrecht, Reidel. (Originally published in *Journal of Philosophy*, 1970.)

Index

This index does not cover the Preface, Bibliography, or section headings. Occasionally a reference that is merely implicit is bracketed.